CAVITATION AND BUBBLE DYNAMICS

Cavitation and Bubble Dynamics deals with the fundamental physical processes of bubble dynamics and the phenomenon of cavitation. It is ideal for graduate students and research engineers and scientists; a basic knowledge of fluid flow and heat transfer is assumed. The analytical methods presented are developed from basic principles.

The book begins with a chapter on nucleation and describes both the theory and observations in flowing and non-flowing systems. Three chapters provide a systematic treatment of the dynamics and growth, collapse, or oscillation of individual bubbles in otherwise quiescent fluids. The following chapters summarize the motion of bubbles in liquids; describe some of the phenomena that occur in homogeneous bubbly flows, with emphasis on cloud cavitation; and summarize some of the experimental observations of cavitating flows. The last chapter provides a review of free streamline methods used to treat separated cavity flows with large attached cavities.

Christopher Earls Brennen is the Richard and Dorothy M. Hayman Professor of Mechanical Engineering Emeritus in the Faculty of Engineering and Applied Science at the California Institute of Technology. He has published more than 200 refereed articles and is especially well known for his research on cavitation, turbomachinery, and multiphase flows. He is the author of three textbooks – *Fundamentals of Multiphase Flows*, *Hydrodynamics of Pumps*, and *Cavitation and Bubble Dynamics* – and has edited several others. Among his many honors and positions are UN Consultant to India (1980); NASA New Technology Award (1980); ASME Centennial Medallion (1980); Chair ASME Fluids Engineering Division (1984–1985); ASME R. T. Knapp Award for Best Paper (1978 and 1981); Caltech Student Award for teaching excellence (1982, 1995, 1996); Christensen Fellow, St. Catherine's College, Oxford (1992); Fellow of the Japan Society for Promotion of Science (1993); JSME Fluids Engineering Award (2002); and Vice-President of Student Affairs at Caltech.

CAVITATION AND BUBBLE DYNAMICS

Christopher Earls Brennen
California Institute of Technology

CAMBRIDGE
UNIVERSITY PRESS

32 Avenue of the Americas, New York NY 10013-2473, USA

Cambridge University Press is part of the University of Cambridge.

It furthers the University's mission by disseminating knowledge in the pursuit of
education, learning and research at the highest international levels of excellence.

www.cambridge.org
Information on this title: www.cambridge.org/9781107644762

First published 2014

A catalogue record for this publication is available from the British Library

Library of Congress Cataloguing in Publication data
Brennen, Christopher E. (Christopher Earls), 1941– author.
Cavitation and bubble dynamics / Christopher E. Brennen, California Institute of Technology.
    pages   cm
Includes bibliographical references and index.
ISBN 978-1-107-64476-2 (pbk.)
1. Bubbles–Dynamics.   I. Title.
QC183.B815 2014
532'.597–dc23   2013022118

ISBN 978-1-107-64476-2 Paperback

"Where Alph, the sacred river, ran
Through caverns measureless to man ..."

Samuel Taylor Coleridge (1772–1834)

Contents

Preface

This book is intended as a combination of a reference book for those who work with cavitation or bubble dynamics and as a monograph for advanced students interested in some of the basic problems associated with this category of multiphase flows. A book like this has many roots. It began many years ago when, as a young post-doctoral Fellow at the California Institute of Technology, I was asked to prepare a series of lectures on cavitation for a graduate course cum seminar series. It was truly a baptism by fire, for the audience included three of the great names in cavitation research, Milton Plesset, Allan Acosta, and Theodore Wu, none of whom readily accepted superficial explanations. For that, I am immensely grateful. The course and I survived, and it evolved into one part of a graduate program in multiphase flows.

There are many people to whom I owe a debt of gratitude for the roles they played in making this book possible. It was my great good fortune to have known and studied with six outstanding scholars, Les Woods, George Gadd, Milton Plesset, Allan Acosta, Ted Wu, and Rolf Sabersky. I benefited immensely from their scholarship and their friendship. I also owe much to my many colleagues in the American Society of Mechanical Engineers whose insights fill many of the pages of this monograph. The support of my research program by the Office of Naval Research is also greatly appreciated. And, of course, I feel honored to have worked with an outstanding group of graduate students at Caltech, including Sheung-Lip Ng, Kiam Oey, David Braisted, Luca d'Agostino, Steven Ceccio, Sanjay Kumar, Douglas Hart, Yan Kuhn de Chizelle, Beth McKenney, Zhenhuan Liu, Yi-Chun Wang, Ronald J. Franz, Tricia A. Waniewski, Fabrizio d'Auria, Garrett E. Reisman, Mark E. Duttweiler, and Keita Ando, all of whom studied aspects of cavitating flows.

The original edition of this book was published by Oxford University Press in 1995. This corrected edition is published by Cambridge University Press and I wish to express my sincerest thanks to Peter Gordon of Cambridge for his expert help and encouragement on this and other projects.

This edition is dedicated with great love and deep gratitude to my wife and lifelong friend, Barbara.

Pasadena, Calif.   C.E.B.
January 2013

Nomenclature

Roman Letters

a	Amplitude of wave-like disturbance
A	Cross-sectional area or cloud radius
b	Body half-width
B	Tunnel half-width
c	Concentration of dissolved gas in liquid, speed of sound, chord
c_k	Phase velocity for wavenumber k
c_P	Specific heat at constant pressure
C_D	Drag coefficient
C_{ij}	Lift/drag coefficient matrix
C_L	Lift coefficient
$\tilde{C}_{Lh}, \tilde{C}_{Lp}$	Unsteady lift coefficients
C_M	Moment coefficient
$\tilde{C}_{Mh}, \tilde{C}_{Mp}$	Unsteady moment coefficients
C_p	Coefficient of pressure
C_{pmin}	Minimum coefficient of pressure
d	Cavity half-width, blade thickness to spacing ratio
D	Mass diffusivity
f	Frequency in Hz
f	Complex velocity potential, $\phi + i\psi$
f_N	A thermodynamic property of the phase or component, N
Fr	Froude number
g	Acceleration due to gravity
g_N	A thermodynamic property of the phase or component N
g_x	Component of the gravitational acceleration in direction x
$\mathcal{G}(f)$	Spectral density function of sound
h	Specific enthalpy, wetted surface elevation, blade tip spacing
H	Henry's law constant
Hm	Haberman-Morton number, normally $g\mu^4/\rho S^3$
i, j, k	Indices

i	Square root of -1 in free streamline analysis
I	Acoustic impulse
I^*	Dimensionless acoustic impulse, $4\pi I \mathcal{R}/\rho_L U_\infty R_H^2$
I_{Ki}	Kelvin impulse vector
j	Square root of -1
k	Boltzmann's constant, polytropic constant or wavenumber
k_N	Thermal conductivity or thermodynamic property of N
K_G	Gas constant
K_{ij}	Added mass coefficient matrix, $M_{ij}/\frac{4}{3}\rho\pi R^3$
Kc	Keulegan-Carpenter number
Kn	Knudsen number, $\lambda/2R$
ℓ	Typical dimension in the flow, cavity half-length
L	Latent heat of vaporization
m	Mass
m_G	Mass of gas in bubble
m_p	Mass of particle
M_{ij}	Added mass matrix
n	Index used for harmonics or number of sites per unit area
$N(R)$	Number density distribution function of R
$\dot{N}_E$	Cavitation event rate
Nu	Nusselt number
p	Pressure
p_a	Radiated acoustic pressure
p_s	Root mean square sound pressure
p_S	A sound pressure level
p_G	Partial pressure of gas
P	Pseudo-pressure
Pe	Peclet number, usually WR/α_L
q	Magnitude of velocity vector
q_c	Free surface velocity
Q	Source strength
r	Radial coordinate
R	Bubble radius
R_B	Equivalent volumetric radius, $[3\tau/4\pi]^{\frac{1}{3}}$
R_H	Headform radius
R_M	Maximum bubble radius
R_N	Cavitation nucleus radius
R_P	Nucleation site radius
$\mathcal{R}$	Distance to measurement point
Re	Reynolds number, usually $2WR/\nu_L$
s	Coordinate measured along a streamline or surface
s	Specific entropy
S	Surface tension

St	Strouhal number, $2fR/W$
t	Time
t_R	Relaxation time for relative motion
t_*	Dimensionless time, t/t_R
T	Temperature
u, v, w	Velocity components in Cartesian coordinates
u_i	Velocity vector
u_r, u_θ	Velocity components in polar coordinates
u'	Perturbation velocity in x direction, $u - U_\infty$
U, U_i	Fluid velocity and velocity vector in absence of particle
U_∞	Velocity of upstream uniform flow
V, V_i	Absolute velocity and velocity vector of particle
w	Complex conjugate velocity, $u - iv$
w	Dimensionless relative velocity, W/W_∞
W	Relative velocity of particle
W_∞	Terminal velocity of particle
We	Weber number, $2\rho W^2 R/S$
z	Complex position vector, $x + iy$

Greek Letters

α	Thermal diffusivity, volume fraction, angle of incidence
β	Cascade stagger angle, other local variables
γ	Ratio of specific heats of gas
Γ	Circulation, other local parameters
δ	Boundary layer thickness or increment of frequency
δ_D	Dissipation coefficient
δ_T	Thermal boundary layer thickness
ϵ	Fractional volume
ζ	Complex variable, $\xi + i\eta$
η	Bubble population per unit liquid volume
η	Coordinate in ζ-plane
θ	Angular coordinate or direction of velocity vector
κ	Bulk modulus of compressibility
λ	Mean free path of molecules or particles
Λ	Accommodation coefficient
μ	Dynamic viscosity
ν	Kinematic viscosity
ξ	Coordinate in ζ-plane
ϖ	Logarithmic hodograph variable, $\chi + i\theta$
ρ	Density
σ	Cavitation number
σ_c	Choked cavitation number

σ_{ij}	Stress tensor
Σ	Thermal parameter in bubble growth
τ	Volume of particle or bubble
ϕ	Velocity potential
ϕ'	Acceleration potential
φ	Fractional perturbation in bubble radius
Φ	Potential energy
χ	$log(q_c/|w|)$
ψ	Stream function
ω	Radian frequency
ω^*	Reduced frequency, $\omega c/U_\infty$

Subscripts

On any variable, Q:

Q_o	Initial value, upstream value, or reservoir value
Q_1, Q_2, Q_3	Components of Q in three Cartesian directions
Q_1, Q_2	Values upstream and downstream of a shock
Q_∞	Value far from the bubble or in the upstream flow
Q_B	Value in the bubble
Q_C	Critical values and values at the critical point
Q_E	Equilibrium value or value on the saturated liquid/vapor line
Q_G	Value for the gas
Q_i	Components of vector Q
Q_{ij}	Components of tensor Q
Q_L	Saturated liquid value
Q_n	Harmonic of order n
Q_P	Peak value
Q_S	Value on the interface or at constant entropy
Q_V	Saturated vapor value
Q_*	Value at the throat

Superscripts and Other Qualifiers

On any variable, Q:

$\bar{Q}$	Mean value of Q or complex conjugate of Q
$\tilde{Q}$	Complex amplitude of oscillating Q
$\dot{Q}$	Time derivative of Q
$\ddot{Q}$	Second time derivative of Q
$\hat{Q}(s)$	Laplace transform of $Q(t)$
$\check{Q}$	Coordinate with origin at image point
Q^+, Q^-	Values of Q on either side of a cut in a complex plane

δQ	Small change in Q
$Re\{Q\}$	Real part of Q
$Im\{Q\}$	Imaginary part of Q

Units

In most of this book, the emphasis is placed on the nondimensional parameters that govern the phenomenon being discussed. However, there are also circumstances in which we shall utilize dimensional thermodynamic and transport properties. In such cases the International System of Units is employed using the basic units of mass (kg), length (m), time (s), and absolute temperature (K); where it is particularly convenient units such as a joule (kg m^2/s^2) are occasionally used.

1 Phase Change, Nucleation, and Cavitation

1.1 Introduction

This first chapter will focus on the mechanisms of formation of two-phase mixtures of vapor and liquid. Particular attention will be given to the process of the creation of vapor bubbles in a liquid. In doing so we will attempt to meld together several overlapping areas of research activity. First, there are the studies of the fundamental physics of nucleation as epitomized by the books of Frenkel (1955) and Skripov (1974). These deal largely with very pure liquids and clean environments in order to isolate the behavior of pure liquids. On the other hand, most engineering systems are impure or contaminated in ways that have important effects on the process of nucleation. The later part of the chapter will deal with the physics of nucleation in such engineering environments. This engineering knowledge tends to be divided into two somewhat separate fields of interest, cavitation and boiling. A rough but useful way of distinguishing these two processes is to define cavitation as the process of nucleation in a liquid when the pressure falls below the vapor pressure, while boiling is the process of nucleation that ocurs when the temperature is raised above the saturated vapor/liquid temperature. Of course, from a basic physical point of view, there is little difference between the two processes, and we shall attempt to review the two processes of nucleation simultaneously. The differences in the two processes occur because of the different complicating factors that occur in a cavitating flow on the one hand and in the temperature gradients and wall effects that occur in boiling on the other hand. The last sections of this first chapter will dwell on some of these complicating factors.

1.2 The Liquid State

Any discussion of the process of phase change from liquid to gas or vice versa must necessarily be preceded by a discussion of the liquid state. Though simple kinetic theory understanding of the gaseous state is sufficient for our purposes, it is necessary to dwell somewhat longer on the nature of the liquid state. In doing so we shall follow Frenkel (1955), though it should also be noted that modern studies are usually couched in terms of statistical mechanics (for example, Carey 1992).

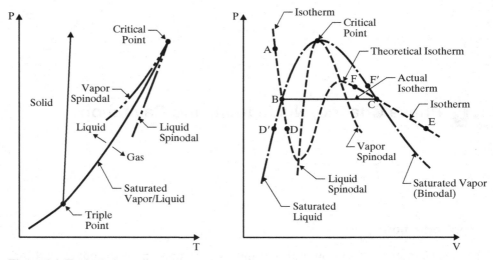

Figure 1.1 Typical phase diagrams.

Our discussion will begin with typical phase diagrams, which, though idealized, are relevant to many practical substances. Figure 1.1 shows typical graphs of pressure, p, temperature, T, and specific volume, V, in which the state of the substance is indicated. The triple point is that point in the phase diagram at which the solid, liquid, and vapor states coexist; that is to say the substance has three alternative stable states. The saturated liquid/vapor line (or binodal) extends from this point to the critical point. Thermodynamically it is defined by the fact that the chemical potentials of the two coexisting phases must be equal. On this line the vapor and liquid states represent two limiting forms of a single "amorphous" state, one of which can be obtained from the other by isothermal volumetric changes, leading through intermediate but unstable states. To quote Frenkel (1955), "Owing to this instability, the actual transition from the liquid state to the gaseous one and vice versa takes place *not* along a *theoretical* isotherm (dashed line, right, Figure 1.1), but along a horizontal isotherm (solid line), corresponding to the splitting up of the original homogeneous substance into two different coexisting phases..." The critical point is that point at which the maxima and minima in the theoretical isotherm vanish and the discontinuity disappears.

The line joining the maxima in the theoretical isotherms is called the vapor spinodal line; the line joining the minima is called the liquid spinodal line Clearly both spinodals end at the critical point. The two regions between the spinodal lines and the saturated (or binodal) lines are of particular interest because the conditions represented by the theoretical isotherm within these regions can be realized in practice under certain special conditions. If, for example, a pure liquid at the state A (Figure 1.1) is depressurized at constant temperature, then several things may happen when the pressure is reduced below that of point B (the saturated vapor pressure). If sufficient numbers of nucleation sites of sufficient size are present (and this needs further discussion later) the liquid will become vapor as the state moves horizontally from B to C, and at pressure below the vapor pressure the state will come to equilibrium in the gaseous region at a point such as E. However, if no nucleation sites are present, the depressurization may lead to continuation of the state down the theoretical isotherm to a point such as D, called a "metastable state" since imperfections may lead to instability and transition to the point E. A liquid at a point such as D is said to be in tension, the pressure difference between B and D being the magnitude of the tension. Of course one could also reach a point like D by proceeding along an isobar from a point such as D' by increasing the temperature. Then an equivalent description of the state at D is to call it superheated and to refer to the difference between the temperatures at D and D' as the superheat.

In an analogous way one can visualize cooling or pressurizing a vapor that is initially at a state such as F and proceeding to a metastable state such as F' where the temperature difference between F and F' is the degree of subcooling of the vapor.

1.3 Fluidity and Elasticity

Before proceding with more detail, it is valuable to point out several qualitative features of the liquid state and to remark on its comparison with the simpler crystalline solid or gaseous states.

The first and most obvious difference between the saturated liquid and saturated vapor states is that the density of the liquid remains relatively constant and similar to that of the solid except close to the critical point. On the other hand the density of the vapor is different by at least 2 and up to 5 or more orders of magnitude, changing radically with temperature. Since it will also be important in later discussions, a plot of the ratio of the saturated liquid density to the saturated vapor density is included as Figure 1.2 for a number of different fluids. The ratio is plotted against a non-dimensional temperature, $\theta = T/T_C$ where T is the actual temperature and T_C is the critical temperature.

Second, an examination of the measured specific heat of the saturated liquid reveals that this is of the same order as the specific heat of the solid except at high temperature close to the critical point. The above two features of liquids imply that the thermal motion of the liquid molecules is similar to that of the solid and involves small amplitude vibrations about a quasi-equilibrium position within the liquid. Thus the arrangement of the molecules has greater similarity with a solid than with a gas. One needs to stress this similarity with a solid to counteract the tendency to think of the liquid state as more akin to the gaseous state than to the solid state because in many observed processes it possesses a dominant fluidity rather than a dominant elasticity. Indeed, it is of interest in this regard to point out that solids also possess fluidity in addition to elasticity. At high temperatures, particularly above 0.6 or 0.7 of the melting temperature, most crystalline solids exhibit a fluidity known as creep. When the strain rate is high, this creep occurs due to the nonisotropic propagation of dislocations (this behavior is not like that of a Newtonian liquid and cannot be characterized by a simple viscosity). At low strain rates, high-temperature creep occurs due simply to the isotropic migration of molecules

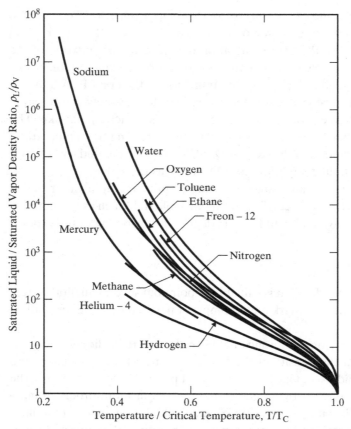

Figure 1.2 Ratio of saturated liquid density to saturated vapor density as a function of temperature for various pure substances.

within the crystal lattice due to the thermal agitation. This kind of creep, which is known as diffusion creep, is analogous to the fluidity observed in most liquids and can be characterized by a simple Newtonian viscosity.

Following this we may ask whether the liquid state possesses an elasticity even though such elasticity may be dominated by the fluidity of the liquid in many physical processes. In both the liquid and solid states one might envisage a certain typical time, t_m, for the migration of a molecule from one position within the structure of the substance to a neighboring position; alternatively one might consider this typical time as characterizing the migration of a "hole" or vacancy from one position to another within the structure. Then if the typical time, t, associated with the applied force is small compared with t_m, the substance will not be capable of permanent deformation during that process and will exhibit elasticity rather than fluidity. On the other hand if $t \gg t_m$ the material will exhibit fluidity. Thus, though the conclusion is overly simplistic, one can characterize a solid as having a large t_m and a liquid as having a small t_m relative to the order of magnitude of the typical time, t, of the applied force. One example of this is that the earth's mantle behaves to all intents and purposes as solid rock in so far as the propagation of seismic waves is concerned, and yet its fluid-like flow over long geological times is responsible for continental drift.

The observation time, t, becomes important when the phenomenon is controlled by stochastic events such as the diffusion of vacancies in diffusion creep. In many cases the process of nucleation is also controlled by such stochastic events, so the observation time will play a significant role in determining this process. Over a longer period of time there is a greater probability that vacancies will coalesce to form a finite vapor pocket leading to nucleation. Conversely, it is also possible to visualize that a liquid could be placed in a state of tension (negative pressure) for a significant period of time before a vapor bubble would form in it. Such a scenario was visualized many years ago. In 1850, Berthelot (1850) subjected purified water to tensions of up to 50 atmospheres before it yielded. This ability of liquids to withstand tension is very similar to the more familiar property exhibited by solids and is a manifestation of the elasticity of a liquid.

1.4 Illustration of Tensile Strength

Frenkel (1955) illustrates the potential tensile strength of a pure liquid by means of a simple, but instructive calculation. Consider two molecules separated by a variable distance s. The typical potential energy, Φ, associated with the intermolecular forces has the form shown in Figure 1.3. Equilibrium occurs at the separation, x_o, typically of the order of $10^{-10} m$. The attractive force, F, between the molecules is equal to $\partial\Phi/\partial x$ and is a maximum at some distance, x_1, where typically x_1/x_o is of the order of 1.1 or 1.2. In a bulk liquid or solid this would correspond to a fractional volumetric expansion, $\Delta V/V_o$, of about one-third. Consequently the application of a constant tensile stress equal to that pertinent at x_1 would completely rupture the liquid or solid since for $x > x_1$ the attractive force is insufficient to counteract that tensile force. In fact, liquids and solids have compressibility moduli, κ, which are usually in the range of 10^{10} to 10^{11} $kg/m s^2$ and since the pressure, $p = -\kappa(\Delta V/V_o)$,

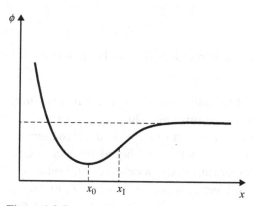

Figure 1.3 Intermolecular potential.

it follows that the typical pressure that will rupture a liquid, p_T, is -3×10^9 to -3×10^{10} $kg/m\ s^2$. In other words, we estimate on this basis that liquids or solids should be able to withstand tensile stresses of 3×10^4 to 3×10^5 atmospheres! In practice solids do not reach these limits (the rupture stress is usually about 100 times less) because of stress concentrations; that is to say, the actual stress encountered at certain points can achieve the large values quoted above at certain points even when the overall or globally averaged stress is still 100 times smaller. In liquids the large theoretical values of the tensile strength defy all practical experience; this discrepancy must be addressed.

It is valuable to continue the above calculation one further step (Frenkel 1955). The elastic energy stored per unit volume of the above system is given by $\kappa(\Delta V)^2/2V_o$ or $|p|\Delta V_o/2$. Consequently the energy that one must provide to pull apart all the molecules and vaporize the liquid can be estimated to be given by $|p_T|/6$ or between 5×10^8 and 5×10^9 $kg/m\ s^2$. This is *in agreement* with the order of magnitude of the latent heat of vaporization measured for many liquids. Moreover, one can correctly estimate the order of magnitude of the critical temperature, T_C, by assuming that, at that point, the kinetic energy of heat motion, kT_C per molecule (where k is Boltzmann's constant, 1.38×10^{-23} $kg\ m^2/s^2 K$) is equal to the energy required to pull all the molecules apart. Taking a typical 10^{30} molecules per m^3, this implies that T_C is given by equating the kinetic energy of the thermal motions per unit volume, or $1.38 \times 10^7 \times T_C$, to $|p_T|/6$. This yields typical values of T_C of the order of $30 \rightarrow 300°K$, which is in accord with the order of magnitude of the actual values. Consequently we find that this simplistic model presents a dilemma because though it correctly predicts the order of magnitude of the latent heat of vaporization and the critical temperature, it fails dismally to predict the tensile strength that a liquid can withstand. One must conclude that unlike the latent heat and critical temperature, the tensile strength is determined by weaknesses at points within the liquid. Such weaknesses are probably ephemeral and difficult to quantify, since they could be caused by minute impurities. This difficulty and the dependence on the time of application of the tension greatly complicate any theoretical evaluation of the tensile strength.

1.5 Cavitation and Boiling

As we discussed in Section 1.2, the tensile strength of a liquid can be manifest in at least two ways:

1. A liquid at constant temperature could be subjected to a decreasing pressure, p, which falls below the saturated vapor pressure, p_V. The value of $(p_V - p)$ is called the tension, Δp, and the magnitude at which rupture occurs is the tensile strength of the liquid, Δp_C. The process of rupturing a liquid by decrease in pressure at roughly constant liquid temperature is often called cavitation.
2. A liquid at constant pressure may be subjected to a temperature, T, in excess of the normal saturation temperature, T_S. The value of $\Delta T = T - T_S$ is the

superheat, and the point at which vapor is formed, ΔT_C, is called the critical superheat. The process of rupturing a liquid by increasing the temperature at roughly constant pressure is often called boiling.

Though the basic mechanics of cavitation and boiling must clearly be similar, it is important to differentiate between the thermodynamic paths that precede the formation of vapor. There are differences in the practical manifestations of the two paths because, although it is fairly easy to cause uniform changes in pressure in a body of liquid, it is very difficult to uniformly change the temperature. Note that the critical values of the tension and superheat may be related when the magnitudes of these quantities are small. By the Clausius-Clapeyron relation,

$$\left(\frac{dp}{dT}\right)_{\substack{saturation \\ conditions}} = \frac{L}{T\left[\rho_V^{-1} - \rho_L^{-1}\right]} \tag{1.1}$$

where ρ_L, ρ_V are the saturated liquid and vapor densities and L is the latent heat of evaporation. Except close to the critical point, we have $\rho_L \gg \rho_V$ and hence $dp/dT \approx \rho_V L/T$. Therefore

$$\Delta T_C \approx \Delta p_C \cdot \frac{T}{L\rho_V} \tag{1.2}$$

For example, in water at $373K$ with $\rho_V = 1\ kg/m^3$ and $L \approx 2 \times 10^6\ m^2/s^2$ a superheat of $20K$ corresponds approximately to one atmosphere of tension. It is important to emphasize that Equation (1.2) is limited to small values of the tension and superheat but provides a useful relation under those circumstances. When Δp_C and ΔT_C are larger, it is necessary to use an appropriate equation of state for the substance in order to establish a numerical relationship.

1.6 Types of Nucleation

In any practical experiment or application weaknesses can typically occur in two forms. The thermal motions within the liquid form temporary, microscopic voids that can constitute the nuclei necessary for rupture and growth to macroscopic bubbles. This is termed homogeneous nucleation. In practical engineering situations it is much commoner to find that the major weaknesses occur at the boundary between the liquid and the solid wall of the container or between the liquid and small particles suspended in the liquid. When rupture occurs at such sites, it is termed heterogeneous nucleation.

In the following sections we briefly review the theory of homogeneous nucleation and some of the experimental results conducted in very clean systems that can be compared with the theory.

In covering the subject of homogeneous nucleation, it is important to remember that the classical treatment using the kinetic theory of liquids allows only weaknesses of one type: the ephemeral voids that happen to occur because of the thermal motions of the molecules. In any real system several other types of weakness are possible. First, it is possible that nucleation might occur at the junction of the liquid and a solid boundary. Kinetic theories have also been developed to cover such heterogeneous nucleation and allow evaluation of whether the chance that this will occur is larger or smaller than the chance of homogeneous nucleation. It is important to remember that heterogeneous nucleation could also occur on very small, sub-micron sized contaminant particles in the liquid; experimentally this would be hard to distinguish from homogeneous nucleation.

Another important form of weaknesses are micron-sized bubbles (microbubbles) of contaminant gas, which could be present in crevices within the solid boundary or within suspended particles or could simply be freely suspended within the liquid. In water, microbubbles of air seem to persist almost indefinitely and are almost impossible to remove completely. As we discuss later, they seem to resist being dissolved completely, perhaps because of contamination of the interface. While it may be possible to remove most of these nuclei from a small research laboratory sample, their presence dominates most engineering applications. In liquids other than water, the kinds of contamination which can occur in practice have not received the same attention.

Another important form of contamination is cosmic radiation. A collision between a high energy particle and a molecule of the liquid can deposit sufficient energy to initiate nucleation when it would otherwise have little chance of occurring. Such, of course, is the principal of the bubble chamber (Skripov 1974). While this subject is beyond the scope of this text, it is important to bear in mind that naturally occurring cosmic radiation could be a factor in promoting nucleation in all of the circumstances considered here.

1.7 Homogeneous Nucleation Theory

Studies of the fundamental physics of the formation of vapor voids in the body of a pure liquid date back to the pioneering work of Gibbs (Gibbs 1961). The modern theory of homogeneous nucleation is due to Volmer and Weber (1926), Farkas (1927), Becker and Doring (1935), Zeldovich (1943), and others. For reviews of the subject, the reader is referred to the books of Frenkel (1955) and Skripov (1974), to the recent text by Carey (1992) and to the reviews by Blake (1949), Bernath (1952), Cole (1970), Blander and Katz (1975), and Lienhard and Karimi (1981). We present here a brief and simplified version of homogeneous nucleation theory, omitting many of the detailed thermodynamical issues; for more detail the reader is referred to the above literature.

In a pure liquid, surface tension is the macroscopic manifestation of the intermolecular forces that tend to hold molecules together and prevent the formation of large holes. The liquid pressure, p, exterior to a bubble of radius R, will be related to the interior pressure, p_B, by

$$p_B - p = \frac{2S}{R} \tag{1.3}$$

where S is the surface tension. In this and the section which follow it is assumed that the concept of surface tension (or, rather, surface energy) can be extended down to bubbles or vacancies a few intermolecular distances in size. Such an approximation is surprisingly accurate (Skripov 1974).

If the temperature, T, is uniform and the bubble contains only vapor, then the interior pressure p_B will be the saturated vapor pressure $p_V(T)$. However, the exterior liquid pressure, $p = p_V - 2S/R$, will have to be less than p_V in order to produce equilibrium conditions. Consequently if the exterior liquid pressure is maintained at a constant value just slightly less than $p_V - 2S/R$, the bubble will grow, R will increase, the excess pressure causing growth will increase, and rupture will occur. It follows that if the maximum size of vacancy present is R_C (termed the critical radius or cluster radius), then the tensile strength of the liquid, Δp_C, will be given by

$$\Delta p_C = 2S/R_C \qquad (1.4)$$

In the case of ephemeral vacancies such as those created by random molecular motions, this simple expression, $\Delta p_C = 2S/R_C$, must be couched in terms of the probability that a vacancy, R_C, will occur during the time for which the tension is applied or the time of observation. This would then yield a probability that the liquid would rupture under a given tension during the available time.

It is of interest to substitute a typical surface tension, $S = 0.05 \ kg/s^2$, and a critical vacancy or bubble size, R_C, comparable with the intermolecular distance of $10^{-10} \ m$. Then the calculated tensile strength, Δp_C, would be $10^9 \ kg/m \ s^2$ or $10^4 \ atm$. This is clearly in accord with the estimate of the tensile strength outlined in Section 1.4 but, of course, at variance with any of the experimental observations.

Equation (1.4) is the first of three basic relations that constitute homogeneous nucleation theory. The second expression we need to identify is that giving the increment of energy that must be deposited in the body of the pure liquid in order to create a nucleus or microbubble of the critical size, R_C. Assuming that the critical nucleus is in thermodynamic equilibrium with its surroundings after its creation, then the increment of energy that must be deposited consists of two parts. First, energy must be deposited to account for that stored in the surface of the bubble. By definition of the surface tension, S, that amount is S per unit surface area for a total of $4\pi R_C^2 S$. But, in addition, the liquid has to be displaced outward in order to create the bubble, and this implies work done on or by the system. The pressure difference involved in this energy increment is the difference between the pressure inside and outside of the bubble (which, in this evaluation, is Δp_C, given by Equation (1.4)). The work done is the volume of the bubble multiplied by this pressure difference, or $4\pi R_C^3 \Delta p_C/3$, and this is the work done by the liquid to achieve the displacement implied by the creation of the bubble. Thus the net energy, W_{CR}, that must be deposited to form the bubble is

$$W_{CR} = 4\pi R_C^2 S - \frac{4}{3}\pi R_C^3 \Delta p_C = \frac{4}{3}\pi R_C^2 S \qquad (1.5)$$

It can also be useful to eliminate R_C from Equations (1.4) and (1.5) to write the expression for the critical deposition energy as

$$W_{CR} = 16\pi S^3 / 3(\Delta p_C)^2 \qquad (1.6)$$

It was, in fact, Gibbs (1961) who first formulated this expression. For more detailed considerations the reader is referred to the works of Skripov (1974) and many others.

The final step in homogeneous nucleation theory is an evaluation of the mechansims by which energy deposition could occur and the probability of that energy reaching the magnitude, W_{CR}, in the available time. Then Equation (1.6) yields the probability of the liquid being able to sustain a tension of Δp_C during that time. In the body of a pure liquid completely isolated from any external radiation, the issue is reduced to an evaluation of the probability that the stochastic nature of the thermal motions of the molecules would lead to a local energy perturbation of magnitude W_{CR}. Most of the homogeneous nucleation theories therefore relate W_{CR} to the typical kinetic energy of the molecules, namely kT (k is Boltzmann's constant) and the relationship is couched in terms of a Gibbs number,

$$Gb = W_{CR}/kT \qquad (1.7)$$

It follows that a given Gibbs number will correspond to a certain probability of a nucleation event in a given volume during a given available time. For later use it is wise to point out that other basic relations for W_{CR} have been proposed. For example, Lienhard and Karimi (1981) find that a value of W_{CR} related to kT_C (where T_C is the critical temperature) rather than kT leads to a better correlation with experimental observations.

A number of expressions have been proposed for the precise form of the relationship between the nucleation rate, J, defined as the number of nucleation events occurring in a unit volume per unit time and the Gibbs number, Gb, but all take the general form

$$J = J_O e^{-Gb} \qquad (1.8)$$

where J_O is some factor of proportionality. Various functional forms have been suggested for J_O. A typical form is that given by Blander and Katz (1975), namely

$$J_O = N \left(\frac{2S}{\pi m} \right)^{\frac{1}{2}} \qquad (1.9)$$

where N is the number density of the liquid (molecules/m^3) and m is the mass of a molecule. Though J_O may be a function of temperature, the effect of an error in J_O is small compared with the effect on the exponent, Gb, in Equation (1.8).

1.8 Comparison with Experiments

The nucleation rate, J, is given by Equations (1.8), (1.7), (1.6), and some form for J_O, such as Equation (1.9). It varies with temperature in ways that are important to identify in order to understand the experimental observations. Consider the tension, Δp_C, which corresponds to a given nucleation rate, J, according to these equations:

$$\Delta p_C = \left[\frac{16 \pi S^3}{3kT \ln (J_O/J)} \right]^{\frac{1}{2}} \qquad (1.10)$$

This can be used to calculate the tensile strength of the liquid given the temperature, T, knowledge of the surface tension variation with temperature, and other fluid properties, plus a selected criterion defining a specific critical nucleation rate, J. Note first that the most important effect of the temperature on the tension occurs through the variation of the S^3 in the numerator. Since S is roughly linear with T declining to zero at the critical point, it follows that Δp_C will be a strong function of temperature close to the critical point because of the S^3 term. In contrast, any temperature dependence of J_O is almost negligible because it occurs in the argument of the logarithm. At lower temperatures, far from the critical point, the dependence of Δp_C on temperature is weak since S^3 varies little, so the tensile strength, Δp_C, will not change much with temperature.

For reasons that will become clear as we progress, it is convenient to divide the discussion of the experimental results into two temperature ranges: above and below that temperature for which the spinodal pressure is roughly zero. This dividing temperature can be derived from an applicable equation of state and turns out to be about $T/T_C = 0.9$.

For temperatures between T_C and $0.9\,T_C$, the tensile strengths calculated from Equation (1.10) are fairly modest. This is because the critical cluster radii, $R_C = 2S/\Delta p_C$, is quite large. For example, a tension of 1 *bar* corresponds to a nucleus $R_C = 1\ \mu m$. It follows that sub-micron-sized contamination particles or microbubbles will have little effect on the experiments in this temperature range because the thermal weaknesses are larger. Figure 1.4, taken from Skripov (1974), presents typical experimental values for the average lifetime, J^{-1}, of a unit volume of superheated liquid, in this case diethyl ether. The data is plotted against the saturation temperature, T_S, for experiments conducted at four different, positive pressures (since the pressures are positive, all the data lies in the $T_C > T > 0.9\,T_C$ domain). Figure 1.4 illustrates several important features. First, all of the data for $J^{-1} < 5s$ correspond to homogeneous nucleation and show fairly good agreement with homogeneous nucleation theory. The radical departure of the experimental data from the theory for $J^{-1} > 5s$ is caused by radiation that induces nucleation at much smaller superheats. The figure also illustrates how weakly the superheat limit depends on the selected value of the "critical" nucleation rate, as was anticipated in our comments on Equation (1.10). Since the lines are almost vertical, one can obtain from the experimental results a maximum possible superheat or tension without the need to stipulate a specific critical nucleation rate. Figure 1.5, taken from Eberhart and Schnyders (1973), presents data on this superheat limit for five different liquids.

For most liquids in this range of positive pressures, the maximum possible superheat is accurately predicted by homogeneous nucleation theory. Indeed,

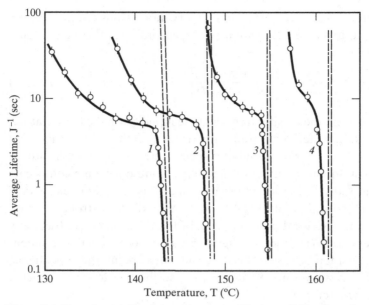

Figure 1.4 Experimentally observed average lifetimes (1/J) of a unit volume of superheated diethyl ether at four different pressures of (1) 1 *bar* (2) 5 *bar* (3) 10 *bar* and (4) 15 *bar* plotted against the saturation temperature, T_S. Lines correspond to two different homogeneous nucleation theories. (From Skripov 1974).

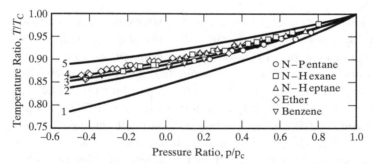

Figure 1.5 Limit of superheat data for five different liquids compared with the liquid spinodal lines derived from five different equations of state including van der Waal's (1) and Berthelot's (5). (From Eberhart and Schnyders 1973).

Lienhard and Karimi (1981) have demonstrated that this limit should be so close to the liquid spinodal line that the data can be used to test model equations of state for the liquid in the metastable region. Figure 1.5 includes a comparison with several such constitutive laws. The data in Figure 1.5 correspond with a critical Gibbs number of 11.5, a value that can be used with Equations (1.6) and (1.7) to yield a simple expression for the superheat limit of most liquids in the range of positive pressures.

Unfortunately, one of the exceptions to the rule is the most common liquid of all, water. Even for $T > 0.9 T_C$, experimental data lie well below the maximum superheat prediction. For example, the estimated temperature of maximum superheat at atmospheric pressure is about 300°C and the maximum that has been attained experimentally is 280°C. The reasons for this discrepancy do not seem to be well understood (Eberhart and Schnyders 1973).

The above remarks addressed the range of temperatures above $0.9\,T_C$. We now turn to the differences that occur at lower temperatures. Below about $0.9T_C$, the superheat limit corresponds to a negative pressure. Indeed, Figure 1.5 includes data down to about $-0.4\,p_C$ ($T \approx 0.85\,T_C$) and demonstrates that the prediction of the superheat limit from homogeneous nucleation theory works quite well down to this temperature. Lienhard and Karimi (1981) have examined the theoretical limit for water at even lower temperatures and conclude that a more accurate criterion than $Gb = 11.5$ is $W_{CR}/kT_C = 11.5$.

One of the reasons for the increasing inaccuracy and uncertainty at lower temperatures is that the homogeneous nucleation theory implies larger and larger tensions, Δp_C, and therefore smaller and smaller critical cluster radii. It follows that almost all of the other nucleation initiators become more important and cause rupture at tensions much smaller than predicted by homogeneous nucleation theory. In water, the uncertainty that was even present for $T > 0.9\,T_C$ is increased even further, and homogeneous nucleation theory becomes virtually irrelevant in water at normal temperatures.

1.9 Experiments on Tensile Strength

Experiments on the tensile strength of water date back to Berthelot (1850) whose basic method has been subsequently used by many investigators. It consists of sealing very pure, degassed liquid in a freshly formed capillary tube under vacuum conditions. Heating the tube causes the liquid to expand, filling the tube at some elevated temperature (and pressure). Upon cooling, rupture is observed at some particular temperature (and pressure). The tensile strength is obtained from these temperatures and assumed values of the compressibility of the liquid. Other techniques used include the mechanical bellows of Vincent (1941) (see also Vincent and Simmonds 1943), the spinning U-tube of Reynolds (1882), and the piston devices of Davies et al. (1956). All these experiments are made difficult by the need to carefully control not only the purity of the liquid but also the properties of the solid surfaces. In many cases it is very difficult to determine whether homogeneous nucleation has occurred or whether the rupture occurred at the solid boundary. Furthermore, the data obtained from such experiments are very scattered.

In freshly drawn capillary tubes, Berthelot (1850) was able to achieve tensions of 50 *bar* in water at normal temperatures. With further refinements, Dixon (1909) was able to get up to 200 *bar* but still, of course, far short of the theoretical limit. Similar scattered results have been reported for water and other liquids by Meyer (1911), Vincent (1941), and others. It is clear that the material of the container plays an important role; using steel Berthelot tubes, Rees and Trevena (1966) were not able to approach the high tensions observed in glass tubes. Clearly, then, the data show that the tensile strength is a function of the contamination of the liquid and the character of the containing surface, and we must move on to consider some of the important issues in this regard.

1.10 Heterogeneous Nucleation

In the case of homogeneous nucleation we considered microscopic voids of radius R, which grow causing rupture when the pressure on the liquid, p, is reduced below the critical value $p_V - 2S/R$. Therefore the tensile strength was $2S/R$. Now consider a number of analogous situations at a solid/liquid interface as indicated in Figure 1.6. The contact angle at the liquid/vapor/solid intersection is denoted by θ. It follows that the tensile strength in the case of the flat hydrophobic surface is given by $2S\sin\theta/R$ where R is the typical maximum dimension of the void. Hence, in theory, the tensile strength could be zero in the limit as $\theta \to \pi$. On the other hand, the tensile strength for a hydrophilic surface is comparable with that for homogeneous nucleation since the maximum dimensions of the voids are comparable. One could therefore conclude that the presence of a hydrophobic surface would cause heterogeneous nucleation and much reduced tensile strength.

Of course, at the microscopic scale with which we are concerned, surfaces are not flat, so we must consider the effects of other local surface geometries. The conical cavity of case (c) is usually considered in order to exemplify the effect of surface geometry. If the half angle at the vertex of this cavity is denoted by α, then it is clear that zero tensile strength occurs at the more realizable value of $\theta = \alpha + \pi/2$ rather than $\theta \to \pi$. Moreover, if $\theta > \alpha + \pi/2$, it is clear that the vapor bubble would grow to fill the cavity at pressures above the vapor pressure.

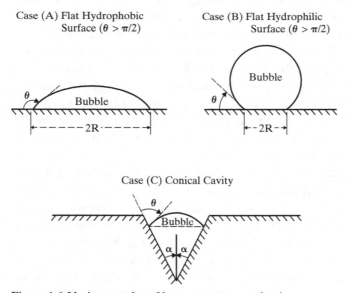

Figure 1.6 Various modes of heterogeneous nucleation.

Hence if one considers the range of microscopic surface geometries, then it is not at all surprising that vapor pockets would grow within some particular surface cavities at pressures in the neighborhood of the vapor pressure, particularly when the surface is hydrophobic. Several questions do however remain. First, how might such a vapor pocket first be created? In most experiments it is quite plausible to conceive of minute pockets of contaminant gas absorbed in the solid surface. This is perhaps least likely with freshly formed glass capillary tubes, a fact that may help explain the larger tensions measured in Berthelot tube experiments. The second question concerns the expansion of these vapor pockets beyond the envelope of the solid surface and into the body of the liquid. One could still argue that dramatic rupture requires the appearance of large voids in the body of the liquid and hence that the flat surface configurations should still be applicable on a larger scale. The answer clearly lies with the detailed topology of the surface. If the opening of the cavity has dimensions of the order of 10^{-5} m, the subsequent tension required to expand the bubble beyond the envelope of the surface is only of the order of a tenth of an atmosphere and hence quite within the realm of experimental observation.

It is clear that some specific sites on a solid surface will have the optimum geometry to promote the growth and macroscopic appearance of vapor bubbles. Such locations are called *nucleation sites*. Furthermore, it is clear that as the pressure is reduced more and more, sites will become capable of generating and releasing bubbles to the body of the liquid. These events are readily observed when you boil a pot of water on the stove. At the initiation of boiling, bubbles are produced at a few specific sites. As the pot gets hotter more and more sites become activated. Hence the density of nucleation sites as a function of the superheat is an important component in the quantification of nucleate boiling.

1.11 Nucleation Site Populations

In pool boiling the hottest liquid is in contact with the solid heated wall of the pool, and hence all the important nucleation sites occur in that surface. For the purpose of quantifying the process of nucleation it is necessary to define a surface number density distribution function for the nucleation sites, $N(R_P)$, where $N(R_P)dR_P$ is the number of sites with size between R_P and $R_P + dR_P$ per unit surface area (thus N has units m^{-3}). In addition to this, it is necessary to know the range of sizes brought into operation by a given superheat, ΔT. Characteristically, all sizes greater than R_P^* will be excited by a tension of $\beta S/R_P^*$ where β is some constant of order unity. This corresponds to a critical superheat given by

$$\Delta T = \beta S T / L\rho_V R_P^* \tag{1.11}$$

Thus the number of sites per unit surface area, $n(\Delta T)$, brought into operation by a specific superheat, ΔT, is given by

$$n(\Delta T) = \int\limits_{\frac{\beta S T}{L\rho_V \Delta T}}^{\infty} N(R_P)\,dR_P \tag{1.12}$$

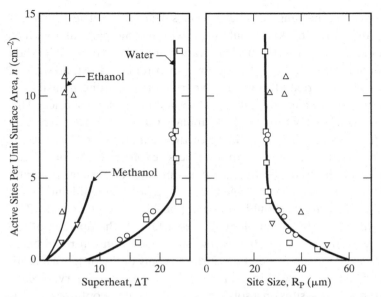

Figure 1.7 Experimental data on the number of active nucleation sites per unit surface area, n, for a polished copper surface. From Griffith and Wallis (1960).

The data of Griffith and Wallis (1960), presented in Figure 1.7, illustrates this effect. On the left of this figure are the measurements of the number of active sites per unit surface area, n, for a particular polished copper surface and the three different liquids. The three curves would correspond to different $N(R_P)$ for the three liquids. The graph on the right is obtained using Equation (1.11) with $\beta = 2$ and demonstrates the veracity of Equation (1.12) for a particular surface.

Identification of the nucleation sites involved in the process of cavitation is much more difficult and has sparked a number of controversies in the past. This is because, unlike pool boiling where the largest tensions are experienced by liquid in contact with a heated surface, a reduction in pressure is experienced by the liquid bulk. Consequently very small particles or microbubbles present as contaminants in the bulk of the liquid are also potential nucleation sites. In particular, cavities in micron-sized particles were first suggested by Harvey et al. (1944) as potential "cavitation nuclei." In the context of cavitating flows such particles are called "free stream nuclei" to distinguish them from the "surface nuclei" present in the macroscopic surfaces bounding the flow. As we shall see later, many of the observations of the onset of cavitation appear to be the result of the excitation of free stream nuclei rather than surface nuclei. Hence there is a need to characterize these free stream nuclei in any particular technological context and a need to control their concentration in any basic experimental study. Neither of these tasks is particularly easy; indeed, it was not until recently that reliable methods for the measurement of free stream nuclei number densities were developed for use in liquid systems of any size. Methods used in the past include the analysis of samples by Coulter counter, and acoustic and light scattering techniques (Billet 1985). However, the most reliable data are probably obtained from holograms of the liquid, which can be reconstructed and microscopically inspected. The resulting size distributions are usually presented as nuclei number density distribution functions, $N(R_N)$, such that the number of free stream nuclei in the size range from R_N to $R_N + dR_N$ present in a unit volume is $N(R_N)dR_N$ (N has units m^{-4}). Illustrated in Figure 1.8 are some typical distributions measured in the filtered and deaerated water of three different water tunnels and in the Pacific Ocean off Los Angeles, California (O'Hern et al. 1985, 1988). Other observations (Billet 1985) produce distributions of similar general shape (roughly $N \propto R_N^{-4}$ for $R_N > 5 \ \mu m$) but with larger values at higher air contents.

It is much more difficult to identify the character of these nuclei. As discussed in the next section, there are real questions as to how small gas-filled microbubbles could exist for any length of time in a body of liquid that is not saturated with that gas. It is not possible to separately assess the number of solid particles and the number of microbubbles with most of the existing experimental techniques. Though both

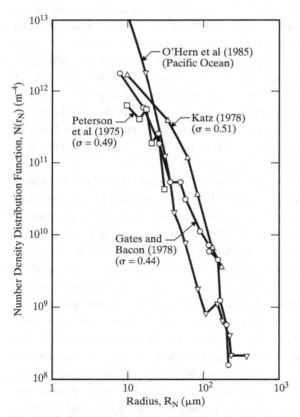

Figure 1.8 Cavitation nuclei number density distribution functions measured by holography in three different water tunnels (Peterson et al. 1975, Gates and Bacon 1978, Katz 1978) at the cavitation numbers, σ, as shown) and in the ocean off Los Angeles, Calif. (O'Hern et al. 1985, 1988).

can act as cavitation nucleation sites, it is clear that microbubbles will more readily grow to observable macroscopic bubbles. One method that has been used to count only those nuclei that will cavitate involves withdrawing sample fluid and sucking it through a very small venturi. Nuclei cavitate at the low pressure in the throat and can be counted provided the concentration is small enough so that the events are separated in time. Then the concentrations of nuclei can be obtained as functions of the pressure level in the throat if the flow rate is known. Such devices are known as cavitation susceptibility meters and tend to be limited to concentrations less than $10 \; cm^{-3}$ (Billet 1985).

If all of the free stream nuclei were uniform in composition and character, one could conclude that a certain tension Δp would activate all nuclei larger than $\beta \Delta p / S$ where β is constant. However, the lack of knowledge of the composition and character of the nuclei as well as other fluid mechanical complications greatly reduces the value of such a statement.

1.12 Effect of Contaminant Gas

Virtually all liquids contain some dissolved gas. Indeed it is virtually impossible to eliminate this gas from any substantial liquid volume. For example, it takes weeks of deaeration to reduce the concentration of air in the water of a tunnel below 3 *ppm* (saturation at atmospheric pressure is about 15 *ppm*). If the nucleation bubble contains some gas, then the pressure in the bubble is the sum of the partial pressure of this gas, p_G, and the vapor pressure. Hence the equilibrium pressure in the liquid is $p = p_V + p_G - 2S/R$ and the critical tension is $2S/R - p_G$. Thus dissolved gas will decrease the potential tensile strength; indeed, if the concentration of gas leads to sufficiently large values of p_G, the tensile strength is negative and the bubble will grow at liquid pressures greater than the vapor pressure.

We refer in the above to circumstances in which the liquid is *not* saturated with gas at the pressure at which it has been stored. In theory, no gas bubbles can exist in equilibrium in a liquid unsaturated with gas but otherwise pure if the pressure is maintained above $p_V + p_G$ where p_G is the equilibrium gas pressure (see Section 2.6). They should dissolve and disappear, thus causing a dramatic increase in the tensile strength of the liquid. While it is true that degassing or high pressure treatment does cause some increase in tensile strength (Keller 1974), the effect is not as great as one would expect. This dilemma has sparked some controversy in the past and at least three plausible explanations have been advanced, all of which have some merit. First is the Harvey nucleus mentioned earlier in which the bubble exists in a crevice in a particle or surface and persists because its geometry is such that the free surface has a highly convex curvature viewed from the fluid so that surface tension supports the high liquid pressure. Second and more esoteric is the possibility of the continuous production of nuclei by cosmic radiation. Third is the proposal by Fox and Herzfeld (1954) of an "organic skin" that gives the free surface of the bubble sufficient elasticity to withstand high pressure. Though originally less plausible than the first two possibilities, this explanation is now more widely accepted because of recent advances in surface rheology, which show that quite small amounts of contaminant in the liquid can generate large elastic surface effects. Such contamination of the surface has also been detected by electron microscopy.

1.13 Nucleation in Flowing Liquids

Perhaps the commonest occurrence of cavitation is in flowing liquid systems where hydrodynamic effects result in regions of the flow where the pressure falls below the vapor pressure. Reynolds (1873) was among the first to attempt to explain the unusual behaviour of ship propellers at the higher rotational speeds that were being achieved during the second half of the ninteenth century. Reynolds focused on the possibility of the entrainment of air into the wakes of the propellor blades, a phenomenon we now term "ventilation." He does not, however, seem to have envisaged the possibility of vapor-filled wakes, and it was left to Parsons (1906) to recognize the role played by vaporization. He also conducted the first experiments on "cavitation" (a word suggested by Froude), and the phenomenon has been a subject of intensive research ever since because of the adverse effects it has on performance, because of the noise it creates and, most surprisingly, the damage it can do to nearby solid surfaces.

For the purposes of the present discussion we shall consider a steady, single-phase flow of a Newtonian liquid of constant density, ρ_L, velocity field, $u_i(x_i)$, and pressure, $p(x_i)$. In all such flows it is convenient to define a reference velocity, U_∞, and reference pressure, p_∞. In external flows around solid bodies, U_∞ and p_∞ are conventionally the velocity and pressure of the uniform, upstream flow. The equations of motion are such that changing the reference pressure results in the same uniform change to the pressure throughout the flow field. Thus the pressure coefficient

$$C_p(x_i) = \frac{p(x_i) - p_\infty}{\frac{1}{2}\rho U_\infty^2} \tag{1.13}$$

is independent of p_∞ for a given geometry of the macroscopic flow boundaries. Furthermore, there will be some location, x_i^*, within the flow where C_p and p are a minimum, and that value of $C_p(x_i^*)$ will be denoted for convenience by C_{pmin}. Note that this is a *negative* number.

Viscous effects within the flow are characterized by the Reynolds number, $Re = \rho_L U_\infty \ell / \mu_L = U_\infty \ell / \nu_L$ where μ_L and ν_L are the dynamic and kinematic viscosities of the liquid and ℓ is the characterized length scale. For a given geometry, $C_p(x_i)$ and C_{pmin} are functions only of Re in steady flows. In the idealized case of an inviscid, frictionless liquid, Bernoulli's equation applies and $C_p(x_i)$ and C_{pmin} become dependent only on the geometry of the flow boundaries and not on any other parameters. For purposes of the present discussion, we shall suppose that for the flow geometry under consideration, the value of C_{pmin} for the single-phase flow is known either from experimental measurement or theoretical calculation.

The stage is therefore set to consider what happens in a given flow when either the overall pressure is decreased or the flow velocity is increased so that the pressure at some point in the flow approaches the vapor pressure, p_V, of the liquid at the reference temperature, T_∞. In order to characterize this relationship, it is conventional to define the *cavitation number, σ* as

$$\sigma = \frac{p_\infty - p_V(T_\infty)}{\frac{1}{2}\rho_L U_\infty^2} \qquad (1.14)$$

Any flow, whether cavitating or not, has some value of σ. Clearly if σ is sufficiently large (p_∞ sufficiently large compared with $p_V(T_\infty)$ or U_∞ sufficiently small), single-phase liquid flow will occur. However, as σ is reduced, nucleation will first occur at some particular value of σ called the incipient cavitation number and denoted by σ_i. For the moment we shall ignore the practical difficulties involved in observing cavitation inception. Further reduction in σ below σ_i causes an increase in the number and extent of vapor bubbles.

In the hypothetical flow of a liquid that cannot withstand any tension and in which vapor bubbles appear instantaneously when p reaches p_V, it is clear that

$$\sigma_i = -C_{pmin} \qquad (1.15)$$

and hence the incipient cavitation number could be ascertained from observations or measurements of the single-phase flow. To exemplify this, consider the nucleation of a free stream nucleus as it travels along the streamline containing x_i^* (see Figure 1.9). For $\sigma > -C_{pmin}$ the pressure along the entire trajectory is greater than p_V. For $\sigma = -C_{pmin}$ the nucleus encounters $p = p_V$ only for an infinitesmal moment. For $\sigma < -C_{pmin}$ the nucleus experiences $p < p_V$ for a finite time. In so far as free steam nuclei are concerned, two factors can cause σ_i to be different from $-C_{pmin}$ (remember again that $-C_{pmin}$ is generally a positive number). First,

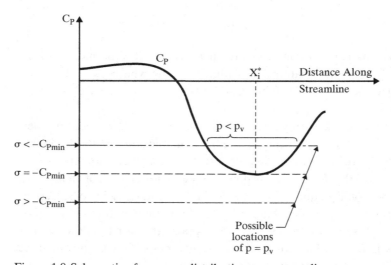

Figure 1.9 Schematic of pressure distribution on a streamline.

nucleation may not occur at $p = p_V$. In a degassed liquid nucleation may require a positive tension, say Δp_C, and hence nucleation would require a cavitation number less than $-C_{pmin}$, namely $\sigma_i = -C_{pmin} - \Delta p_C/\frac{1}{2}\rho_L U_\infty^2$. In a liquid containing a great deal of contaminant gas Δp_C could actually be negative, so that σ_i would be larger than $-C_{pmin}$. Second, growth of a nucleus to a finite, observable size requires a finite time under conditions $p < p_V - \Delta p_C$. This residence time effect will cause the observed σ_i to be less than $-C_{pmin} - \Delta p_C/\frac{1}{2}\rho_L U_\infty^2$. As we shall see in the next chapter, the rate of growth of a bubble can also be radically affected by the thermodynamic properties of the liquid and vapor which are, in turn, functions of the temperature of the liquid. Consequently σ_i may also depend on the liquid temperature.

1.14 Viscous Effects in Cavitation Inception

The discussion in the previous section was deliberately confined to ideal, steady flows. When the flow is also assumed to be inviscid, the value of $-C_{pmin}$ is a simple positive constant for a given flow geometry. However, when the effects of viscosity are included, C_{pmin} will be a function of Reynolds number, Re, and even in a steady flow one would therefore expect to observe a dependence of the incipient cavitation number, σ_i, on the Reynolds number. For convenience, we shall refer to this as the steady viscous effect.

Up to this point we have assumed that the flow and the pressures are laminar and steady. However, most of the flows with which the engineer must deal are not only turbulent but also unsteady. Vortices occur not only because they are inherent in turbulence but also because of both free and forced shedding of vortices. This has important consequences for cavitation inception because the pressure in the center of a vortex may be significantly lower than the mean pressure in the flow. The measurement or calculation of $-C_{pmin}$ would elicit the value of the lowest mean pressure, while cavitation might first occur in a transient vortex whose core pressure was much lower than the lowest mean pressure. Unlike the residence time factor, this would tend to cause higher values of σ_i than would otherwise be expected. It would also cause σ_i to change with Reynolds number, Re. Note that this would be separate from the effect of Re on C_{pmin} and, to distinguish it, we shall refer to it as the turbulence effect.

In summary, there are a number of reasons for σ_i to be different from the value of $-C_{pmin}$ that might be calculated from knowledge of the pressures in the single-phase liquid flow:

1. Existence of a tensile strength can cause a reduction in σ_i.
2. Residence time effects can cause a reduction in σ_i.
3. Existence of contaminant gas can cause an increase in σ_i.
4. Steady viscous effect due to dependence of C_{pmin} on Re can cause σ_i to be a function of Re.
5. Turbulence effects can cause an increase in σ_i.

If it were not for these effects, the prediction of cavitation would be a straight-forward matter of determining C_{pmin}. Unfortunately, these effects can cause large departures from the criterion, $\sigma_i = -C_{pmin}$, with important engineering consequences in many applications.

Furthermore, the above discussion identifies the parameters that must be controlled or at least measured in systematic experiments on cavitation inception:

1. The cavitation number, σ.
2. The Reynolds number, Re.
3. The liquid temperature, T_∞.
4. The liquid quality, including the number and nature of the free stream nuclei, the amount of dissolved gas, and the free stream turbulence.
5. The quality of the solid, bounding surfaces, including the roughness (since this may affect the hydrodynamics) and the porosity or pit population.

Since this is a tall order, and many of the effects such as the interaction of turbulence and cavitation inception have only recently been identified, it is not surprising that the individual effects are not readily isolated from many of the experiments performed in the past. Nevertheless, some discussion of these experiments is important for practical reasons.

1.15 Cavitation Inception Measurements

To illustrate some of the effects described in the preceding section, we shall attempt to give a brief overview of the extensive literature on the subject of cavitation inception. For more detail, the reader is referred to the reviews by Acosta and Parkin (1975), Arakeri (1979), and Rood (1991), as well as to the book by Knapp, Daily, and Hammitt (1970).

The first effect that we illustrate is that of the uncertainty in the tensile strength of the liquid. It is very difficult to characterize and almost impossible to remove from a substantial body of liquid (such as that used in a water tunnel) all the particles, microbubbles, and contaminant gas that will affect nucleation. This can cause substantial differences in the cavitation inception numbers (and, indeed, the form of cavitation) from different facilities and even in the same facility with differently treated water. The ITTC (International Towing Tank Conference) comparative tests (Lindgren and Johnsson 1966, Johnsson 1969) provided a particularly dramatic example of these differences when cavitation on the same axisymmetric headform (called the ITTC headform) was examined in many different water tunnels around the world. An example of the variation of σ_i in those experiments is reproduced as Figure 1.10.

As a further illustration, Figure 1.11 reproduces data obtained by Keller (1974) for the cavitation inception number in flows around hemispherical bodies. The water was treated in different ways so that it contained different populations of nuclei as shown on the left in Figure 1.11. As one might anticipate, the water with the higher nuclei population had a substantially larger cavitation inception number.

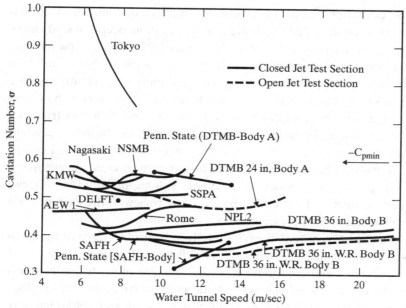

Figure 1.10 The inception numbers measured for the same axisymmetric headform in a variety of water tunnels around the world. Data collected as part of a comparative study of cavitation inception by the International Towing Tank Conference (Lindgren and Johnsson 1966, Johnsson 1969).

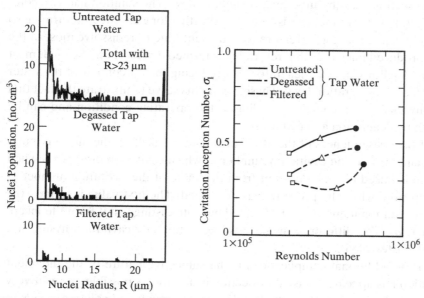

Figure 1.11 Histograms of nuclei populations in treated and untreated tap water and the corresponding cavitation inception numbers on hemispherical headforms of three different diameters, 3 cm (●), 4.5 cm (△), and 6 cm (□) (Keller 1974).

Because the cavitation nuclei are crucial to an understanding of cavitation inception, it is now recognized that the liquid in any cavitation inception study must be monitored by measuring the number of nuclei present in the liquid. Typical nuclei number distributions from water tunnels and from the ocean were shown earlier in Figure 1.8. It should, however, be noted that most of the methods currently used for making these measurements are still in the development stage. Devices based on acoustic scattering and on light scattering have been explored. Other instruments known as cavitation susceptibility meters cause samples of the liquid to cavitate and measure the number and size of the resulting macroscopic bubbles. Perhaps the most reliable method has been the use of holography to create a magnified three-dimensional photographic image of a sample volume of liquid, which can then be surveyed for nuclei. Billet (1985) has recently reviewed the current state of cavitation nuclei measurements (see also Katz et al. 1984).

It may be of interest to note that cavitation itself is also a source of nuclei in many facilities. This is because air dissolved in the liquid will tend to come out of solution at low pressures and contribute a partial pressure of air to the contents of any macroscopic cavitation bubble. When that bubble is convected into regions of higher pressure and the vapor condenses, this leaves a small air bubble that only redissolves very slowly, if at all. This unforeseen phenomenon caused great trauma for the first water tunnels, which were modeled directly on wind tunnels. It was discovered that after a few minutes of operating with a cavitating body in the working section, the bubbles produced by the cavitation grew rapidly in number and began to complete the circuit of the facility to return in the incoming flow. Soon the working section was obscured by a two-phase flow. The solution had two components. First, a water tunnel needs to be fitted with a long and deep return leg so that the water remains at high pressure for sufficient time to redissolve most of the cavitation-produced nuclei. Such a return leg is termed a "resorber." Second, most water tunnel facilities have a "deaerator" for reducing the air content of the water to 20 to 50% of the saturation level. These comments serve to illustrate the fact that $N(R_N)$ in any facility can change according to the operating condition and can be altered both by deaeration and by filtration.

One of the consequences of the effect of cavitation itself on the nuclei population in a facility is that the cavitation number at which cavitation disappears when the pressure is raised may be different from the value of the cavitation number at which it appeared when the pressure was decreased. The first value is termed the "desinent" cavitation number and is denoted by σ_d to distinguish it from the inception number, σ_i. The difference in these values is termed "cavitation hysteresis" (Holl and Treaster 1966).

One of the additional complications is the subjective nature of the judgment that cavitation has appeared. Visual inspection is not always possible, nor is it very objective since the number of events (single bubble growth and collapse) tends to increase gradually over a range of cavitation numbers. If, therefore, one made a judgment based on a certain event rate, it is inevitable that the inception cavitation number would increase with nuclei population. Experiments have found that the production of noise is a simpler and more repeatable measure of inception than visual observation. While still subject to the variations with nuclei population discussed above, it has the advantage of being quantifiable.

Most of the data of Figure 1.8 is taken from water tunnel water that has been somewhat filtered and degassed or from the ocean, which is surprisingly clean. Thus there are very few nuclei with a size greater than 100 μm. On the other hand, there are many hydraulic applications in which the water contains much larger gas bubbles. These can then grow substantially as they pass through a region of low pressure in the pump or other hydraulic device, even though the pressure is everywhere above the vapor pressure. Such a phenomenon is called "pseudo-cavitation." Though a cavitation inception number is not particularly relevant to such circumstances, attempts to measure σ_i under these circumstances would clearly yield values much larger than $-C_{pmin}$.

On the other hand, if the liquid is quite clean with only very small nuclei, the tension that this liquid can sustain would mean that the minimum pressure would have to fall well below p_V for inception to occur. Then σ_i would be much smaller than $-C_{pmin}$. Thus it is clear that the quality of the water and its nuclei could cause the cavitation inception number to be either larger or smaller than $-C_{pmin}$.

1.16 Cavitation Inception Data

Though much of the inception data in the literature is deficient in the sense that the nuclei population and character are unknown, it is nevertheless of value to review some of the important trends in that data base. In doing so we could be reassured that each investigator probably applied a consistent criterion in assessing cavitation inception. Therefore, though the data from different investigators and facilities may be widely scattered, one would hope that the trends exhibited in a particular research project would be qualitatively significant.

Consider first the inception characteristics of a single hydrofoil as the angle of attack is varied. The data of Kermeen (1956), obtained for a NACA 4412 hydrofoil, is reproduced in Figure 1.12. At positive angles of attack the regions of low pressure and cavitation inception will occur on the suction surface; at negative angles of attack these phenomena will shift to the pressure surface. Furthermore, as the angle of attack is increased in either direction, the value of $-C_{pmin}$ will increase, and hence the inception cavitation number will also increase.

As we will discuss in the next section, the scaling of cavitation inception with changes in the size and speed of the hydraulic device can be an important issue, particularly when scaling the results from model-scale water tunnel experiments to prototypes as is necessary, for example, in developing ship propellers. Typical data on cavitation inception for a single hydrofoil (Holl and Wislicenus 1961) is reproduced in Figure 1.13. Data for three different sizes of 12% Joukowski hydrofoil (at zero angle of attack) were obtained at different speeds. They were plotted against Reynolds number in the hope that this would reduce the data to a single curve. The fact that this did not occur demonstrates that there is a size or speed effect

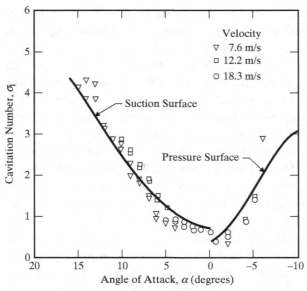

Figure 1.12 Cavitation inception characteristics of a NACA 4412 hydrofoil (Kermeen 1956).

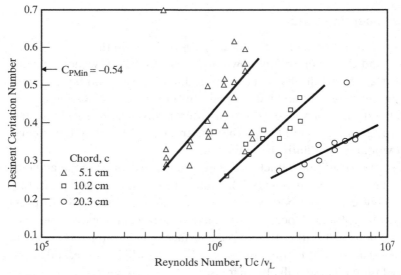

Figure 1.13 The desinent cavitation numbers for three sizes of Joukowski hydrofoils at zero angle of attack and as a function of Reynolds number, *Re* (Holl and Wislicenus 1961). Note the theoretical $C_{pmin} = -0.54$.

separate from that due to the Reynolds number. It seems reasonable to suggest that the missing parameter is the ratio of the nuclei size to chord length; however, in the absence of information on the nuclei, such conclusions are purely speculative.

To complete the list of those factors that may influence cavitation inception, it is necessary to mention the effects of surface roughness and of the turbulence level in the flow. The two effects are connected to some degree since roughness will affect the level of turbulence. But roughness can also affect the flow by delaying boundary layer separation and therefore affecting the pressure and velocity fields in a more global manner. The reader is referred to Arndt and Ippen (1968) for details of the effects of surface roughness on cavitation inception.

Turbulence affects cavitation inception since a nucleus may find itself in the core of a vortex where the pressure level is lower than the mean. It could therefore cavitate when it might not do so under the influence of the mean pressure level. Thus turbulence may promote cavitation, but one must allow for the fact that it may alter the global pressure field by altering the location of flow separation. These complicated viscous effects on cavitation inception were first examined in detail by Arakeri and Acosta (1974) and Gates and Acosta (1978) (see also Arakeri 1979). The implications for cavitation inception in the highly turbulent environment of many internal flows such as occur in pumps have yet to be examined in detail.

1.17 Scaling of Cavitation Inception

The complexity of the issues raised in the last section helps to explain why serious questions remain as to how to scale cavitation inception. This is perhaps one of the most troublesome issues a hydraulic engineer must face. Model tests of a ship's propeller or large pump-turbine may allow the designer to accurately estimate the noncavitating performance of the device. However, he will not be able to place anything like the same confidence in his ability to scale the cavitation inception data.

Consider the problem in more detail. Changing the size of the device will alter not only the residence time effect but also the Reynolds number. Furthermore, the nuclei will now be a different size relative to the device than in the model. Changing the speed in an attempt to maintain Reynolds number scaling may only confuse the issue by further alterating the residence time. Moreover, changing the speed will also change the cavitation number. To recover the modeled condition, one must then change the pressure level, which may alter the nuclei content. There is also the issue of what to do about the surface roughness in the model and in the prototype.

The other issue of scaling that arises is how to anticipate the cavitation phenomena in one liquid based on data obtained in another. It is clearly the case that the literature contains a great deal of data on water. Data on other liquids are quite meager. Indeed, I have not located any nuclei number distributions for a fluid other than water. Since the nuclei play such a key role, it is not surprising that our current ability to scale from one liquid to another is quite tentative.

It would not be appropriate to leave this subject without emphasizing that most of the remarks in the last two sections have focused on the inception of cavitation. Once cavitation has become established, the phenomena that occur are much less sensitive to special factors such as the nuclei content. Hence the scaling of developed cavitation can proceed with much more confidence than the scaling of cavitation inception. This is not, however, of much solace to the engineer charged with avoiding cavitation completely.

References

Acosta, A.J. and Parkin, B.R. (1975). Cavitation inception—a selective review. *J. Ship Res.*, **19**, 193–205.

Arakeri, V.H. (1979). Cavitation inception. *Proc. Indian Acad. Sci.*, **C2**, Part 2, 149–177.

Arakeri, V.H. and Acosta, A.J. (1974). Viscous effects in the inception of cavitation on axisymmetric bodies. *ASME J. Fluids Eng.*, **95**, No. 4, 519–528.

Arndt, R.E.A. and Ippen, A.T. (1968). Rough surface effects on cavitation inception. *ASME J. Basic Eng.*, **90**, 249–261.

Becker, R. and Doring, W. (1935). The kinetic treatment of nuclear formation in supersaturated vapors. *Ann. Phys.*, **24**, 719 and 752.

Bernath, L. (1952). Theory of bubble formation in liquids. *Ind. Eng. Chem.*, **44**, No. 6, 1310–1313.

Berthelot, M. (1850). Sur quelques phénomenes de dilation forcée de liquides. *Ann. de Chimie et de Physique*, **30**, 232–237.

Billet, M.L. (1985). Cavitation nuclei measurement—a review. *Proc. 1985 ASME Cavitation and Multiphase Flow Forum*, 31–38.

Blake, F.G. (1949). The tensile strength of liquids; a review of the literature. *Harvard Acou. Res. Lab. Rep. TM9*.

Blander, M. and Katz, J.L. (1975). Bubble nucleation in liquids. *AIChE Journal*, **21**, No. 5, 833–848.

Carey, V.P. (1992). *Liquid-vapor phase-change phenomena*. Hemisphere Publ. Co.

Cole, R. (1970). Boiling nucleation. *Adv. Heat Transfer*, **10**, 86–166.

Davies, R.M., Trevena, D.H., Rees, N.J.M., and Lewis, G.M. (1956). The tensile strength of liquids under dynamic stressing. *Proc. N.P.L. Symp. on Cavitation in Hydrodynamics*.

Dixon, H.H. (1909). Note on the tensile strength of water. *Sci. Proc. Royal Dublin Soc.*, **12**, (N.S.), 60 (see also **14**, (N.S.), 229, (1914)).

Eberhart, J.G. and Schnyders, M.C. (1973). Application of the mechanical stability condition to the prediction of the limit of superheat for normal alkanes, ether, and water. *J. Phys. Chem.*, **77**, No. 23, 2730–2736.

Farkas, L. (1927). The velocity of nucleus formation in supersaturated vapors. *J. Physik Chem.*, **125**, 236.

Fox, F.E. and Herzfeld, K.F. (1954). Gas bubbles with organic skin as cavitation nuclei. *J. Acoust. Soc. Am.*, **26**, 984–989.

Frenkel, J. (1955). *Kinetic theory of liquids*. Dover, New York.

Gates, E.M. and Acosta, A.J. (1978). Some effects of several free stream factors on cavitation inception on axisymmetric bodies. *Proc. 12th Naval Hydrodyn. Symp., Wash. D.C.*, 86–108.

Gates, E.M. and Bacon, J. (1978). Determination of cavitation nuclei distribution by holography. *J. Ship Res.*, **22**, No. 1, 29–31.

Gibbs, W. (1961). *The Scientific Papers, Vol. 1*. Dover Publ. Inc., NY.

Griffith, P. and Wallis, J.D. (1960). The role of surface conditions in nucleate boiling. *Chem. Eng. Prog. Symp.*, Ser. 56, **30**, 49.

Harvey, E.N., Barnes, D.K., McElroy, W.D., Whiteley, A.H., Pease, D.C., and Cooper, K.W. (1944). Bubble formation in animals. I, Physical factors. *J. Cell. and Comp. Physiol.*, **24**, No. 1, 1–22.

Holl, J.W. and Wislicenus, G.F. (1961). Scale effects on cavitation. *ASME J. Basic Eng.*, **83**, 385–398.

Holl, J.W. and Treaster, A.L. (1966). Cavitation hysteresis. *ASME J. Basic Eng.*, **88**, 199–212.

Johnsson, C.A. (1969). Cavitation inception on headforms, further tests. *Proc. 12th Int. Towing Tank Conf., Rome*, 381–392.

Katz, J. (1978). Determination of solid nuclei and bubble distributions in water by holography. *Calif.Inst. of Tech., Eng. and Appl. Sci. Div. Rep. No. 183–3*.

Katz, J., Gowing, S., O'Hern, T., and Acosta, A.J. (1984). A comparitive study between holographic and light-scattering techniques of microbubble detection. *Proc. IUTAM Symp. on Measuring Techniques in Gas-Liquid Two-Phase Flows*, 41–66.

Keller, A.P. (1974). Investigations concerning scale effects of the inception of cavitation. *Proc. I.Mech.E. Conf. on Cavitation*, 109–117.

Kermeen, R.W. (1956). Water tunnel tests of NACA 4412 and Walchner profile 7 hydrofoils in non-Cavitating and cavitating Flows. *Calif. Inst. of Tech. Hydro. Lab. Rep. 47–5*.

Knapp, R.T., Daily, J.W., and Hammitt, F.G. (1970). *Cavitation*. McGraw-Hill, New York.

Lienhard, J.H. and Karimi, A. (1981). Homogeneous nucleation and the spinodal line. *ASME J. Heat Transfer*, **103**, 61–64.

Lindgren, H. and Johnsson, C.A. (1966). Cavitation inception on headforms, ITTC comparitive experiments. *Proc. 11th Towing Tank Conf. Tokyo*, 219–232.

Meyer, J. (1911). Zur Kenntnis des negativen Druckes in Flüssigkeiten. *Abhandl. Dent. Bunsen Ges.*, **III**, No. 1; also No. 6.

O'Hern, T.J., Katz, J., and Acosta, A.J. (1985). Holographic measurements of cavitation nuclei in the sea. *Proc. ASME Cavitation and Multiphase Flow Forum*, 39–42.

O'Hern, T., d'Agostino, L., and Acosta, A.J. (1988). Comparison of holographic and Coulter counter measurements of cavitation nuclei in the ocean. *ASME J. Fluids Eng.*, **110**, 200–207.

Parsons, C.A. (1906). The steam turbine on land and at sea. *Lecture to the Royal Institution, London*.

Peterson, F.B., Danel, F., Keller, A.P., and Lecoffre, Y. (1975). Comparitive measurements of bubble and particulate spectra by three optical methods. *Proc. 14th Int. Towing Tank Conf.*

Rees, E.P. and Trevena, D.H. (1966). Cavitation thresholds in liquids under static conditions. *Proc. ASME Cavitation Forum*, 12 (see also (1967), 1).

Reynolds, O. (1873). The causes of the racing of the engines of screw steamers investigated theoretically and by experiment. *Trans. Inst. Naval Arch.*, **14**, 56–67.

Reynolds, O. (1882). On the internal cohesion of liquids and the suspension of a column of mercury to a height of more than double that of the barometer. *Mem. Manchester Lit. Phil. Soc.*, **7**, 3rd Series, 1.

Rood, E.P. (1991). Mechanisms of cavitation inception—review. *ASME J. Fluids Eng.*, **113**, 163–175.

Skripov, V.P. (1974). *Metastable Liquids*. John Wiley and Sons.

Vincent, R.S. (1941). The measurement of tension in liquids by means of a metal bellows. *Proc. Phys. Soc. (London)*, **53**, 126–140.

Vincent, R.S. and Simmonds, G.H. (1943). Examination of the Berthelot method of measuring tension in liquids. *Proc. Phys. Soc. (London)*, **55**, 376–382.

Volmer, M. and Weber, A. (1926). Keimbildung in übersättigten Gebilden. *Zeit. Physik. Chemie*, **119**, 277–301.

Zeldovich, J.B. (1943). On the theory of new phase formation: cavitation. *Acta Physicochimica, URSS*, **18**, 1–22.

2 Spherical Bubble Dynamics

2.1 Introduction

Having considered the initial formation of bubbles, we now proceed to identify the subsequent dynamics of bubble growth and collapse. The behavior of a single bubble in an infinite domain of liquid at rest far from the bubble and with uniform temperature far from the bubble will be examined first. This spherically symmetric situation provides a simple case that is amenable to analysis and reveals a number of important phenomena. Complications such as those introduced by the presence of nearby solid boundaries will be discussed in the chapters which follow.

2.2 Rayleigh-Plesset Equation

Consider a spherical bubble of radius, $R(t)$ (where t is time), in an infinite domain of liquid whose temperature and pressure far from the bubble are T_∞ and $p_\infty(t)$ respectively. The temperature, T_∞, is assumed to be a simple constant since temperature gradients were eliminated *a priori* and uniform heating of the liquid due to internal heat sources or radiation will not be considered. On the other hand, the pressure, $p_\infty(t)$, is assumed to be a known (and perhaps controlled) input which regulates the growth or collapse of the bubble.

Though compressibility of the liquid can be important in the context of bubble collapse, it will, for the present, be assumed that the liquid density, ρ_L, is a constant. Furthermore, the dynamic viscosity, μ_L, is assumed constant and uniform. It will also be assumed that the contents of the bubble are homogeneous and that the temperature, $T_B(t)$, and pressure, $p_B(t)$, within the bubble are always uniform. These assumptions may not be justified in circumstances that will be identified as the analysis proceeds.

The radius of the bubble, $R(t)$, will be one of the primary results of the analysis. As indicated in Figure 2.1, radial position within the liquid will be denoted by the distance, r, from the center of the bubble; the pressure, $p(r,t)$, radial outward velocity, $u(r,t)$, and temperature, $T(r,t)$, within the liquid will be so designated. Conservation of mass requires that

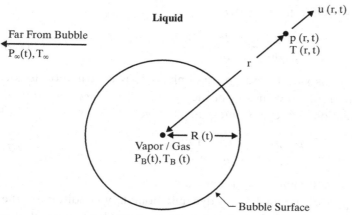

Figure 2.1 Schematic of a spherical bubble in an infinite liquid.

$$u(r,t) = \frac{F(t)}{r^2} \tag{2.1}$$

where $F(t)$ is related to $R(t)$ by a kinematic boundary condition at the bubble surface. In the idealized case of zero mass transport across this interface, it is clear that $u(R,t) = dR/dt$ and hence

$$F(t) = R^2 \frac{dR}{dt} \tag{2.2}$$

But this is often a good approximation even when evaporation or condensation is occurring at the interface. To demonstrate this, consider a vapor bubble. The volume rate of production of vapor must be equal to the rate of increase of size of the bubble, $4\pi R^2 dR/dt$, and therefore the mass rate of evaporation must be $\rho_V(T_B)4\pi R^2 dR/dt$ where $\rho_V(T_B)$ is the saturated vapor density at the bubble temperature, T_B. This, in turn, must equal the mass flow of liquid inward relative to the interface, and hence the inward velocity of liquid relative to the interface is given by $\rho_V(T_B)(dR/dt)/\rho_L$. Therefore

$$u(R,t) = \frac{dR}{dt} - \frac{\rho_V(T_B)}{\rho_L}\frac{dR}{dt} = \left[1 - \frac{\rho_V(T_B)}{\rho_L}\right]\frac{dR}{dt} \tag{2.3}$$

and

$$F(t) = \left[1 - \frac{\rho_V(T_B)}{\rho_L}\right]R^2\frac{dR}{dt} \tag{2.4}$$

In many practical cases $\rho_V(T_B) \ll \rho_L$ and therefore the approximate form of Equation (2.2) may be adequate. For clarity we will continue with the approximate form given in Equation (2.2).

Assuming a Newtonian liquid, the Navier-Stokes equation for motion in the r direction,

$$-\frac{1}{\rho_L}\frac{\partial p}{\partial r} = \frac{\partial u}{\partial t} + u\frac{\partial u}{\partial r} - \nu_L\left[\frac{1}{r^2}\frac{\partial}{\partial r}\left(r^2\frac{\partial u}{\partial r}\right) - \frac{2u}{r^2}\right] \tag{2.5}$$

yields, after substituting for u from $u = F(t)/r^2$:

$$-\frac{1}{\rho_L}\frac{\partial p}{\partial r} = \frac{1}{r^2}\frac{dF}{dt} - \frac{2F^2}{r^5} \qquad (2.6)$$

Note that the viscous terms vanish; indeed, the only viscous contribution to the Rayleigh-Plesset Equation (2.10) comes from the dynamic boundary condition at the bubble surface. Equation (2.6) can be integrated to give

$$\frac{p - p_\infty}{\rho_L} = \frac{1}{r}\frac{dF}{dt} - \frac{1}{2}\frac{F^2}{r^4} \qquad (2.7)$$

after application of the condition $p \to p_\infty$ as $r \to \infty$.

To complete this part of the analysis, a dynamic boundary condition on the bubble surface must be constructed. For this purpose consider a control volume consisting of a small, infinitely thin lamina containing a segment of interface (Figure 2.2). The net force on this lamina in the radially outward direction per unit area is

$$(\sigma_{rr})_{r=R} + p_B - \frac{2S}{R} \qquad (2.8)$$

or, since $\sigma_{rr} = -p + 2\mu_L \partial u/\partial r$, the force per unit area is

$$p_B - (p)_{r=R} - \frac{4\mu_L}{R}\frac{dR}{dt} - \frac{2S}{R} \qquad (2.9)$$

In the absence of mass transport across the boundary (evaporation or condensation) this force must be zero, and substitution of the value for $(p)_{r=R}$ from Equation (2.7) with $F = R^2\frac{dR}{dt}$ yields the generalized Rayleigh-Plesset equation for bubble dynamics:

$$\frac{p_B(t) - p_\infty(t)}{\rho_L} = R\frac{d^2R}{dt^2} + \frac{3}{2}\left(\frac{dR}{dt}\right)^2 + \frac{4\nu_L}{R}\frac{dR}{dt} + \frac{2S}{\rho_L R} \qquad (2.10)$$

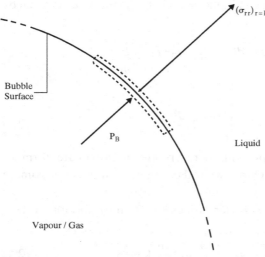

Figure 2.2 Portion of the spherical bubble surface.

Given $p_\infty(t)$ this represents an equation that can be solved to find $R(t)$ provided $p_B(t)$ is known. In the absence of the surface tension and viscous terms, it was first derived and used by Rayleigh (1917). Plesset (1949) first applied the equation to the problem of traveling cavitation bubbles.

2.3 Bubble Contents

In addition to the Rayleigh-Plesset equation, considerations of the bubble contents are necessary. To be fairly general, it is assumed that the bubble contains some quantity of contaminant gas whose partial pressure is p_{Go} at some reference size, R_o, and temperature, T_∞. Then, if there is no appreciable mass transfer of gas to or from the liquid, it follows that

$$p_B(t) = p_V(T_B) + p_{Go} \left(\frac{T_B}{T_\infty} \right) \left(\frac{R_o}{R} \right)^3 \tag{2.11}$$

In some cases this last assumption is not justified, and it is necessary to solve a mass transport problem for the liquid in a manner similar to that used for heat diffusion in the next section (see Section 2.6).

It remains to determine $T_B(t)$. This is not always necessary since, under some conditions, the difference between the unknown T_B and the known T_∞ is negligible. But there are also circumstances in which the temperature difference, $(T_B(t) - T_\infty)$, is important and the effects caused by this difference dominate the bubble dynamics. Clearly the temperature difference, $(T_B(t) - T_\infty)$, leads to a different vapor pressure, $p_V(T_B)$, than would occur in the absence of such thermal effects, and this alters the growth or collapse rate of the bubble. It is therefore instructive to substitute Equation (2.11) into (2.10) and thereby write the Rayleigh-Plesset equation in the following general form:

$$\underset{(1)}{\underbrace{\frac{p_V(T_\infty) - p_\infty(t)}{\rho_L}}} + \underset{(2)}{\underbrace{\frac{p_V(T_B) - p_V(T_\infty)}{\rho_L}}} + \underset{(3)}{\underbrace{\frac{p_{Go}}{\rho_L} \left(\frac{T_B}{T_\infty} \right) \left(\frac{R_o}{R} \right)^3}}$$

$$= \underset{(4)}{\underbrace{R \frac{d^2 R}{dt^2}}} + \frac{3}{2} \left(\frac{dR}{dt} \right)^2 + \underset{(5)}{\underbrace{\frac{4 \nu_L}{R} \frac{dR}{dt}}} + \underset{(6)}{\underbrace{\frac{2S}{\rho_L R}}} \tag{2.12}$$

The first term, (1), is the instantaneous tension or driving term determined by the conditions far from the bubble. The second term, (2), will be referred to as the thermal term, and it will be seen that very different bubble dynamics can be expected depending on the magnitude of this term. When the temperature difference is small, it is convenient to use a Taylor expansion in which only the first derivative is retained to evaluate

$$\frac{p_V(T_B) - p_V(T_\infty)}{\rho_L} = A(T_B - T_\infty) \tag{2.13}$$

where the quantity A may be evaluated from

$$A = \frac{1}{\rho_L}\frac{dp_V}{dT} = \frac{\rho_V(T_\infty)L(T_\infty)}{\rho_L T_\infty} \qquad (2.14)$$

using the Clausius-Clapeyron relation. It is consistent with the Taylor expansion approximation to evaluate ρ_V and L at the known temperature T_∞. It follows that, for small temperature differences, term (2) in Equation (2.12) is given by $A(T_B - T_\infty)$.

The degree to which the bubble temperature, T_B, departs from the remote liquid temperature, T_∞, can have a major effect on the bubble dynamics, and it is neccessary to discuss how this departure might be evaluated. The determination of $(T_B - T_\infty)$ requires two steps. First, it requires the solution of the heat diffusion equation,

$$\frac{\partial T}{\partial t} + \frac{dR}{dt}\left(\frac{R}{r}\right)^2\frac{\partial T}{\partial r} = \frac{\alpha_L}{r^2}\frac{\partial}{\partial r}\left(r^2\frac{\partial T}{\partial r}\right) \qquad (2.15)$$

to determine the temperature distribution, $T(r,t)$, within the liquid (α_L is the thermal diffusivity of the liquid). Second, it requires an energy balance for the bubble. The heat supplied to the interface from the liquid is

$$4\pi R^2 k_L \left(\frac{\partial T}{\partial r}\right)_{r=R} \qquad (2.16)$$

where k_L is the thermal conductivity of the liquid. Assuming that all of this is used for vaporization of the liquid (this neglects the heat used for heating or cooling the existing bubble contents, which is negligible in many cases), one can evaluate the mass rate of production of vapor and relate it to the known rate of increase the volume of the bubble. This yields

$$\frac{dR}{dt} = \frac{k_L}{\rho_V L}\left(\frac{\partial T}{\partial r}\right)_{r=R} \qquad (2.17)$$

where k_L, ρ_V, L should be evaluated at $T = T_B$. If, however, $T_B - T_\infty$ is small, it is consistent with the linear analysis described earlier to evaluate these properties at $T = T_\infty$.

The nature of the thermal effect problem is now clear. The thermal term in the Rayleigh-Plesset Equation (2.12) requires a relation between $(T_B(t) - T_\infty)$ and $R(t)$. The energy balance Equation (2.17) yields a relation between $(\partial T/\partial r)_{r=R}$ and $R(t)$. The final relation between $(\partial T/\partial r)_{r=R}$ and $(T_B(t) - T_\infty)$ requires the solution of the heat diffusion equation. It is this last step that causes considerable difficulty due to the evident nonlinearities in the heat diffusion equation; no exact analytic solution exists. However, the solution of Plesset and Zwick (1952) provides a useful approximation for many purposes. This solution is confined to cases in which the thickness of the thermal boundary layer, δ_T, surrounding the bubble is small compared with the radius of the bubble, a restriction that can be roughly represented by the identity

$$R \gg \delta_T \approx (T_\infty - T_B)/\left(\frac{\partial T}{\partial r}\right)_{r=R} \qquad (2.18)$$

The Plesset-Zwick result is that

$$T_\infty - T_B(t) = \left(\frac{\alpha_L}{\pi}\right)^{\frac{1}{2}} \int_0^t \frac{[R(x)]^2 \left(\frac{\partial T}{\partial r}\right)_{r=R(x)} dx}{\left[\int_x^t [R(y)]^4 dy\right]^{\frac{1}{2}}} \tag{2.19}$$

where x and y are dummy time variables. Using Equation (2.17) this can be written as

$$T_\infty - T_B(t) = \frac{L\rho_V}{\rho_L c_{PL} \alpha_L^{\frac{1}{2}}} \left(\frac{1}{\pi}\right)^{\frac{1}{2}} \int_0^t \frac{[R(x)]^2 \frac{dR}{dt} dx}{[\int_x^t R^4(y) dy]^{\frac{1}{2}}} \tag{2.20}$$

This can be directly substituted into the Rayleigh-Plesset equation to generate a complicated integro-differential equation for $R(t)$. However, for present purposes it is more instructive to confine our attention to regimes of bubble growth or collapse that can be approximated by the relation

$$R = R^* t^n \tag{2.21}$$

where R^* and n are constants. Then the Equation (2.20) reduces to

$$T_\infty - T_B(t) = \frac{L\rho_V}{\rho_L c_{PL} \alpha_L^{\frac{1}{2}}} R^* t^{n-\frac{1}{2}} C(n) \tag{2.22}$$

where the constant

$$C(n) = n \left(\frac{4n+1}{\pi}\right)^{\frac{1}{2}} \int_0^1 \frac{z^{3n-1} dz}{(1 - z^{4n+1})^{\frac{1}{2}}} \tag{2.23}$$

and is of order unity for most values of n of practical interest ($0 < n < 1$ in the case of bubble growth). Under these conditions the linearized thermal term, (2), in the Rayleigh-Plesset Equation (2.12) becomes

$$(T_B - T_\infty) \frac{\rho_V L}{\rho_L T_\infty} = -\Sigma(T_\infty) C(n) R^* t^{n-\frac{1}{2}} \tag{2.24}$$

where the thermodynamic parameter

$$\Sigma(T_\infty) = \frac{L^2 \rho_V^2}{\rho_L^2 c_{PL} T_\infty \alpha_L^{\frac{1}{2}}} \tag{2.25}$$

It will be seen that this parameter, Σ, whose units are $m/sec^{\frac{3}{2}}$, is crucially important in determining the bubble dynamic behavior.

2.4 In the Absence of Thermal Effects

First we consider some of the characteristics of bubble dynamics in the absence of any significant thermal effects. This kind of bubble dynamic behavior is termed "inertially controlled" to distinguish it from the "thermally controlled" behavior discussed later. Under these circumstances the temperature in the liquid is assumed uniform and term (2) in the Rayleigh-Plesset Equation (2.12) is zero.

Furthermore, it will be assumed that the behavior of the gas in the bubble is polytropic so that

$$p_G = p_{Go} \left(\frac{R_o}{R} \right)^{3k} \tag{2.26}$$

where k is approximately constant. Clearly $k = 1$ implies a constant bubble temperature and $k = \gamma$ would model adiabatic behavior. It should be understood that accurate evaluation of the behavior of the gas in the bubble requires the solution of the mass, momentum, and energy equations for the bubble contents combined with appropriate boundary conditions which will include a thermal boundary condition at the bubble wall. Such an analysis would probably assume spherical symmetry. However, it is appropriate to observe that any non-spherically symmetric internal motion would tend to mix the contents and, perhaps, improve the validity of the polytropic assumption.

With the above assumptions the Rayleigh-Plesset equation becomes

$$\frac{p_V(T_\infty) - p_\infty(t)}{\rho_L} + \frac{p_{Go}}{\rho_L} \left(\frac{R_o}{R} \right)^{3k} = R\ddot{R} + \frac{3}{2}(\dot{R})^2 + \frac{4\nu_L \dot{R}}{R} + \frac{2S}{\rho_L R} \tag{2.27}$$

where the overdot denotes d/dt. Equation (2.27) without the viscous term was first derived and used by Noltingk and Neppiras (1950, 1951); the viscous term was investigated first by Poritsky (1952).

Equation (2.27) can be readily integrated numerically to find $R(t)$ given the input $p_\infty(t)$, the temperature T_∞, and the other constants. Initial conditions are also required and, in the context of cavitating flows, it is appropriate to assume that the microbubble of radius R_o is in equilibrium at $t = 0$ in the fluid at a pressure $p_\infty(0)$ so that

$$p_{Go} = p_\infty(0) - p_V(T_\infty) + \frac{2S}{R_o} \tag{2.28}$$

and that $dR/dt|_{t=0} = 0$. A typical solution for Equation (2.27) under these conditions and with a pressure $p_\infty(t)$, which first decreases below $p_\infty(0)$ and then recovers to its original value, is shown in Figure 2.3. The general features of this solution are characteristic of the response of a bubble as it passes through any low-pressure region; they also reflect the strong nonlinearity of Equation (2.27). The growth is fairly smooth and the maximum size occurs after the minimum pressure. The collapse process is quite different. The bubble collapses catastrophically, and this is followed by successive rebounds and collapses. In the absence of dissipation mechanisms such as viscosity these rebounds would continue indefinitely without attenuation.

Analytic solutions to Equation (2.27) are limited to the case of a step function change in p_∞. Nevertheless, these solutions reveal some of the characteristics of more general pressure histories, $p_\infty(t)$, and are therefore valuable to document. With a constant value of $p_\infty(t > 0) = p_\infty^*$, Equation (2.27) is integrated by multiplying through by $2R^2 \dot{R}$ and forming time derivatives. Only the viscous term cannot

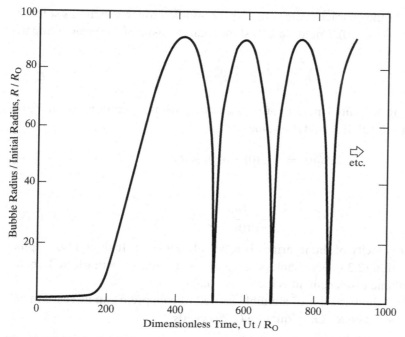

Figure 2.3 Typical solution of the Rayleigh-Plesset equation for spherical bubble size/ initial size, R/R_0. The nucleus enters a low-pressure region at a dimensionless time of 0 and is convected back to the original pressure at a dimensionless time of 500. The low-pressure region is sinusoidal and symmetric about 250.

be integrated in this way, and what follows is confined to the inviscid case. After integration, application of the initial condition $\dot{R}(0) = 0$ yields

$$(\dot{R})^2 = \frac{2(p_V - p_\infty^*)}{3\rho_L}\left[1 - \frac{R_o^3}{R^3}\right] + \frac{2p_{Go}}{3\rho_L(1-k)}\left[\frac{R_o^{3k}}{R^{3k}} - \frac{R_o^3}{R^3}\right] - \frac{2S}{\rho_L R}\left[1 - \frac{R_o^2}{R^2}\right] \qquad (2.29)$$

where, in the case of isothermal gas behavior, the term involving p_{Go} becomes

$$2\frac{p_{Go}}{\rho_L}\frac{R_o^3}{R^3}\ln\left(\frac{R_o}{R}\right) \qquad (2.30)$$

By rearranging Equation (2.29) it follows that

$$t = R_o \int\limits_{1}^{R/R_o}\left[\frac{2\left(p_V - p_\infty^*\right)\left(1 - x^{-3}\right)}{3\rho_L} + \frac{2p_{Go}\left(x^{-3k} - x^{-3}\right)}{3(1-k)\rho_L} - \frac{2S\left(1 - x^{-2}\right)}{\rho_L R_o x}\right]^{-\frac{1}{2}}dx \qquad (2.31)$$

where, in the case $k = 1$, the gas term is replaced by

$$\frac{2p_{Go}}{x^3}\ln x \qquad (2.32)$$

This integral can be evaluated numerically to find $R(t)$, albeit indirectly.

Consider first the characteristic behavior for bubble growth which this solution exhibits when $p_\infty^* < p_\infty(0)$. Equation 2.29 shows that the asymptotic growth rate for $R \gg R_o$ is given by

$$\dot{R} \to \left[\frac{2}{3} \frac{(p_V - p_\infty^*)}{\rho_L} \right]^{\frac{1}{2}} \tag{2.33}$$

Hence, following an initial period of acceleration, whose duration, t_A, may be estimated from this relation and the value of

$$\ddot{R}(0) = (p_\infty(0) - p_\infty^*)/\rho_L R_o \tag{2.34}$$

to be

$$t_A = \left[\frac{2\rho_L R_o^2 (p_V - p_\infty^*)}{3(p_\infty(0) - p_\infty^*)^2} \right]^{\frac{1}{2}} \tag{2.35}$$

the subsequent velocity of the interface is relatively constant. It should be emphasized that Equation (2.33) nevertheless represents explosive growth of the bubble, in which the volume displacement is increasing like t^3.

Now contrast the behavior of a bubble caused to collapse by an increase in p_∞ to p_∞^*. In this case when $R \ll R_o$ Equation (2.29) yields

$$\dot{R} \to -\left(\frac{R_o}{R} \right)^{\frac{3}{2}} \left[\frac{2(p_\infty^* - p_V)}{3\rho_L} + \frac{2S}{\rho_L R_o} - \frac{2p_{Go}}{3(k-1)\rho_L} \left(\frac{R_o}{R} \right)^{3(k-1)} \right]^{\frac{1}{2}} \tag{2.36}$$

where, in the case of $k = 1$, the gas term is replaced by $2p_{Go} \ln(R_o/R)/\rho_L$. However, most bubble collapse motions become so rapid that the gas behavior is much closer to adiabatic than isothermal, and we will therefore assume $k \neq 1$.

For a bubble with a substantial gas content the asymptotic collapse velocity given by Equation (2.36) will not be reached and the bubble will simply oscillate about a new, but smaller, equilibrium radius. On the other hand, when the bubble contains very little gas, the inward velocity will continually increase (like $R^{-3/2}$) until the last term within the square brackets reaches a magnitude comparable with the other terms. The collapse velocity will then decrease and a minimum size given by

$$R_{min} = R_o \left[\frac{1}{(k-1)} \frac{p_{Go}}{(p_\infty^* - p_V + 3S/R_o)} \right]^{\frac{1}{3(k-1)}} \tag{2.37}$$

will be reached, following which the bubble will rebound. Note that, if p_{Go} is small, the R_{min} could be very small indeed. The pressure and temperature of the gas in the bubble at the minimum radius are then given by p_{max} and T_{max} where

$$p_{max} = p_{Go} \left[(k-1)(p_\infty^* - p_V + 3S/R_o)/p_{Go} \right]^{k/(k-1)} \tag{2.38}$$

$$T_{max} = T_o \left[(k-1)(p_\infty^* - p_V + 3S/R_o)/p_{Go} \right] \tag{2.39}$$

We will comment later on the magnitudes of these temperatures and pressures (see Section 3.2).

The case of zero gas content presents a special albeit somewhat hypothetical problem, since apparently the bubble will reach zero size and at that time have an infinite inward velocity. In the absence of both surface tension and gas content, Rayleigh (1917) was able to integrate Equation (2.31) to obtain the time, t_{TC}, required for total collapse from $R = R_o$ to $R = 0$:

$$t_{TC} = 0.915 \left(\frac{\rho_L R_o^2}{p_\infty^* - p_V} \right)^{\frac{1}{2}} \tag{2.40}$$

It is important at this point to emphasize that while the above results for bubble growth are quite practical, the results for bubble collapse may be quite misleading. Apart from the neglect of thermal effects, the analysis was based on two other assumptions that may be violated during collapse. Later we shall see that the final stages of collapse may involve such high velocities (and pressures) that the assumption of liquid incompressibility is no longer appropriate. But, perhaps more important, it transpires (see Chapter 5) that a collapsing bubble loses its spherical symmetry in ways that can have important engineering consequences.

2.5 Stability of Vapor/Gas Bubbles

Apart from the characteristic bubble growth and collapse processes discussed in the last section, it is also important to recognize that the equilibrium condition

$$p_V - p_\infty + p_{GE} - \frac{2S}{R_E} = 0 \tag{2.41}$$

may not always represent a *stable* equilibrium state at $R = R_E$ with a partial pressure of gas p_{GE}.

Consider a small perturbation in the size of the bubble from $R = R_E$ to $R = R_E(1 + \epsilon)$, $\epsilon \ll 1$ and the response resulting from the Rayleigh-Plesset equation. Care must be taken to distinguish two possible cases:

(i) The partial pressure of the gas remains the same at p_{GE}.
(ii) The mass of gas in the bubble and its temperature, T_B, remain the same.

From a practical point of view the Case (i) perturbation is generated over a length of time sufficient to allow adequate mass diffusion in the liquid so that the partial pressure of gas is maintained at the value appropriate to the concentration of gas dissolved in the liquid. On the other hand, Case (ii) is considered to take place too rapidly for significant gas diffusion. It follows that in Case (i) the gas term in the Rayleigh-Plesset Equation (2.27) is p_{GE}/ρ_L whereas in Case (ii) it is $p_{GE} R_E^{3k}/\rho_L R^{3k}$. If n is defined as zero for Case (i) and $n = 1$ for Case (ii) then substitution of $R = R_E(1 + \epsilon)$ into the Rayleigh-Plesset equation yields

$$R\ddot{R} + \frac{3}{2}(\dot{R})^2 + 4\nu_L \frac{\dot{R}}{R} = \frac{\epsilon}{\rho_L} \left[\frac{2S}{R_E} - 3nkp_{GE} \right] \tag{2.42}$$

Note that the right-hand side has the same sign as ϵ if

$$\frac{2S}{R_E} > 3nkp_{GE} \tag{2.43}$$

and a different sign if the reverse holds. Therefore, if the above inequality holds, the left-hand side of Equation (2.42) implies that the velocity and/or acceleration of the bubble radius has the same sign as the perturbation, and hence the equilibrium is *unstable* since the resulting motion will cause the bubble to deviate further from $R = R_E$. On the other hand, the equilibrium is stable if $np_{GE} > 2S/3R_E$.

First consider Case (i) which must always be *unstable* since the inequality (2.43) always holds if $n = 0$. This is simply a restatement of the fact (discussed in Section 2.6) that, if one allows time for mass diffusion, then all bubbles will either grow or shrink indefinitely.

Case (ii) is more interesting since in many of the practical engineering situations pressure levels change over a period of time that is short compared with the time required for significant gas diffusion. In this case a bubble in stable equilibrium requires

$$p_{GE} = \frac{m_G T_B K_G}{\frac{4}{3}\pi R_E^3} > \frac{2S}{3kR_E} \tag{2.44}$$

where m_G is the mass of gas in the bubble and K_G is the gas constant. Indeed for a *given* mass of gas there exists a critical bubble size, R_C, where

$$R_C = \left[\frac{9km_G T_B K_G}{8\pi S} \right]^{1/2} \tag{2.45}$$

This critical radius was first identified by Blake (1949) and Neppiras and Noltingk (1951) and is often referred to as the Blake critical radius. All bubbles of radius $R_E < R_C$ can exist in stable equilibrium, whereas all bubbles of radius $R_E > R_C$ must be unstable. This critical size could be reached by decreasing the ambient pressure from p_∞ to the critical value, $p_{\infty c}$, where from Equations (2.45) and (2.41) it follows that

$$p_{\infty c} = p_V - \frac{4S}{3} \left[\frac{8\pi S}{9km_G T_B K_G} \right]^{\frac{1}{2}} \tag{2.46}$$

which is often called the Blake threshold pressure.

The isothermal case ($k = 1$) is presented graphically in Figure 2.4 where the solid lines represent equilibrium conditions for a bubble of size R_E plotted against the tension ($p_V - p_\infty$) for various fixed masses of gas in the bubble and a fixed surface tension. The critical radius for any particular m_G corresponds to the maximum in each curve. The locus of the peaks is the graph of R_C values and is shown by the dashed line whose equation is $(p_V - p_\infty) = 4S/3R_E$. The region to the right of the dashed line represents unstable equilibrium conditions. This graphical representation was used by Daily and Johnson (1956) and is useful in visualizing the quasistatic response of a bubble when subjected to a decreasing pressure. Starting in the fourth quadrant under conditions in which the ambient pressure $p_\infty > p_V$, and assuming the

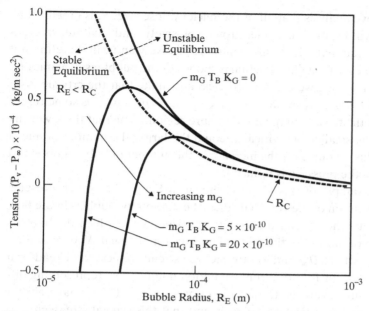

Figure 2.4 Stable and unstable bubble equilibrium radii as a function of the tension for various masses of gas in the bubble. Stable and unstable conditions are separated by the dotted line. Adapted from Daily and Johnson (1956).

mass of gas in the bubble is constant, the radius R_E will first increase as $(p_V - p_\infty)$ increases. The bubble will pass through a series of stable equilibrium states until the particular critical pressure corresponding to the maximum is reached. Any slight decrease in p_∞ below the value corresponding to this point will result in explosive cavitation growth regardless of whether p_∞ is further decreased or not. Indeed, it is clear from this analysis that the critical tension for a liquid should be given by $4S/3R$ rather than $2S/R$ as maintained in Chapter 1, since stable equilibrium conditions do not exist in the range

$$\frac{4S}{3R} < (p_V - p_\infty) < \frac{2S}{R} \qquad (2.47)$$

Other questions arise from inspection of Figure 2.4. Note that for a given subcritical tension two alternate equilibrium states exist, one smaller stable state and one larger unstable state. Suppose that a bubble at the smaller stable state is also subjected to pressure oscillations of sufficient magnitude to cause the bubble to momentarily exceed the size, R_C. It would then grow explosively without bound. This effect is important in understanding the role of turbulence in cavitation inception or the response of a liquid to an acoustic field (see Chapter 4).

This stability phenomenon has important consequences in many cavitating flows. To recognize this, one must visualize a spectrum of sizes of cavitation nuclei being convected into a region of low pressure within the flow. Then the p_∞ in Equations (2.41) and (2.47) will be the local pressure in the liquid surrounding the bubble, and p_∞ must be less than p_V for explosive cavitation growth to occur. It is clear from the above analysis that all of the nuclei whose size, R, is greater than some critical value will become unstable, grow explosively, and cavitate, whereas those nuclei smaller than that critical size will react passively and will therefore not become visible to the eye. Though the actual response of the bubble is dynamic and p_∞ is changing continuously, we can nevertheless anticipate that the crtical nuclei size will be given approximately by $4S/3(p_V - p_\infty)^*$ where $(p_V - p_\infty)^*$ is some representative measure of the tension in the low-pressure region. Note that the lower the pressure level, p_∞, the smaller the critical size and the larger the number of nuclei that are activated. This accounts for the increase in the number of bubbles observed in a cavitating flow as the pressure is reduced.

A quantitative example of this effect is shown in Figure 2.5, which presents results from the integration of the Rayleigh-Plesset equation for bubbles in the flow around an axisymmetric headform. It shows the maximum size which the bubbles achieve as a function of the size of the original nucleus for a typical Weber number, $\rho_L R_H U_\infty^2/S$, of 28,000 where U_∞ and R_H are the free stream velocity and headform radius. Data are plotted for four different cavitation numbers, σ, representing different ambient pressure levels. Note that the curves for $\sigma < 0.5$ all have abrupt vertical sections at certain critical nuclei sizes and that this critical size decreases with decreasing σ. Numerical results for this and other flows show that the critical size, R_C, adheres fairly closely to the nondimensional version of the expression derived earlier,

$$R_C \approx \kappa S/\rho_L U_\infty^2 (-\sigma - C_{pmin})$$ (2.48)

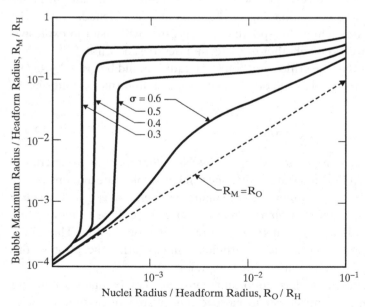

Figure 2.5 The maximum size, R_M, to which a cavitation bubble grows according to the Rayleigh-Plesset equation as a function of the original nucleus size, R_o, and the cavitation number, σ, in the flow around an axisymmetric headform of radius, R_H, with Weber number, $\rho_L R_H U_\infty^2/S = 28,000$ (from Ceccio and Brennen 1991).

where C_{pmin} is the minimum pressure coefficient in the flow and the factor κ is close to unity.

Note also from Figure 2.5 that, whatever their initial size, all unstable nuclei grow to roughly the same maximum size. This is because both the asymptotic growth rate and the time available for growth are relatively independent of the size of the original nucleus. From Equation (2.33) the growth rate is given approximately by

$$\frac{dR}{dt} = U_\infty(-\sigma - C_{pmin})^{\frac{1}{2}} \qquad (2.49)$$

Moreover, if the pressure near the minimum pressure point is represented by

$$C_p = C_{pmin} + C_{p*}(s/R_H)^2 \qquad (2.50)$$

where s is a coordinate measured along the surface, R_H is the typical dimension of the body, and C_{p*} is a constant which is typically of order one, then the typical time available for growth, t_G, is given approximately by

$$t_G = \frac{2R_H(-\sigma - C_{pmin})^{\frac{1}{2}}}{C_{p*}^{\frac{1}{2}} U_\infty(1 - C_{pmin})^{\frac{1}{2}}} \qquad (2.51)$$

It follows that the maximum size, R_M, will be given roughly by

$$\frac{R_M}{R_H} = \frac{2(-\sigma - C_{pmin})}{C_{p*}^{\frac{1}{2}}(1 - C_{pmin})^{\frac{1}{2}}} \qquad (2.52)$$

and therefore only changes modestly with cavitation number within the range of significance.

2.6 Growth by Mass Diffusion

In most of the circumstances considered in this chapter, it is assumed that the events occur too rapidly for significant mass transfer of contaminant gas to occur between the bubble and the liquid. Thus we assumed in Section 2.3 and elsewhere that the mass of contaminant gas in the bubble remained constant. It is convenient to reconsider this issue at this point, for the methods of analysis of mass diffusion will clearly be similar to those of thermal diffusion (Scriven 1959). Moreover, there are some issues that require analysis of the rate of increase or decrease of the mass of gas in the bubble. One of the most basic issues is the fact that any and all of the gas-filled microbubbles that are present in a subsaturated liquid (and particularly in water) should dissolve away if the ambient pressure is sufficiently high. Henry's law states that the partial pressure of gas, p_{GE}, in the bubble, which is in equilibrium with a saturated concentration, c_∞, of gas dissolved in the liquid will be given by

$$p_{GE} = Hc_\infty \qquad (2.53)$$

where H is Henry's law constant for that gas and liquid combination. (Note that H decreases substantially with temperature.) Consequently, if the ambient pressure,

p_∞, is greater than $(Hc_\infty + p_V - 2S/R)$, the bubble should dissolve away completely. As we discussed in Section 1.12, experience is contrary to this theory, and microbubbles persist even when the liquid is subjected to several atmospheres of pressure for an extended period.

The process of mass transfer can be analysed by noting that the concentration, $c(r,t)$, of gas in the liquid will be governed by a diffusion equation identical in form to Equation (2.15),

$$\frac{\partial c}{\partial t} + \frac{dR}{dt}\left(\frac{R}{r}\right)^2 \frac{\partial c}{\partial r} = \frac{D}{r^2}\frac{\partial}{\partial r}\left(r^2\frac{\partial c}{\partial r}\right) \tag{2.54}$$

where D is the mass diffusivity, typically $2 \times 10^{-5}\ cm^2/sec$ for air in water at normal temperatures. As Plesset and Prosperetti (1977) demonstrate, the typical bubble growth rates due to mass diffusion are so slow that the convection term (the second term on the left-hand side of Equation (2.54)) is negligible.

The simplest problem is that of a bubble of radius, R, in a liquid at a fixed ambient pressure, p_∞, and gas concentration, c_∞. In the absence of inertial effects the partial pressure of gas in the bubble will be p_{GE} where

$$p_{GE} = p_\infty - p_V + 2S/R \tag{2.55}$$

and therefore the concentration of gas at the liquid interface is $c_S = p_{GE}/H$. Epstein and Plesset (1950) found an approximate solution to the problem of a bubble in a liquid initially at uniform gas concentration, c_∞, at time, $t = 0$, which takes the form

$$R\frac{dR}{dt} = \frac{D}{\rho_G}\frac{\left[c_\infty - c_S(1 + 2S/Rp_\infty)\right]}{(1 + 4S/3Rp_\infty)}\left[1 + R(\pi Dt)^{-\frac{1}{2}}\right] \tag{2.56}$$

where ρ_G is the density of gas in the bubble and c_S is the saturated concentration at the interface at the partial pressure given by Equation (2.55) (the vapor pressure is neglected in their analysis). The last term in Equation (2.56), $R(\pi Dt)^{-\frac{1}{2}}$, arises from a growing diffusion boundary layer in the liquid at the bubble surface. This layer grows like $(Dt)^{\frac{1}{2}}$. When t is large, the last term in Equation (2.56) becomes small and the characteristic growth is given approximately by

$$[R(t)]^2 - [R(0)]^2 \approx \frac{2D(c_\infty - c_S)t}{\rho_G} \tag{2.57}$$

where, for simplicity, we have neglected surface tension.

It is instructive to evaluate the typical duration of growth (or shrinkage). From Equation (2.57) the time required for complete solution is t_{CS} where

$$t_{CS} \approx \frac{\rho_G[R(0)]^2}{2D(c_S - c_\infty)} \tag{2.58}$$

Typical values of $(c_S - c_\infty)/\rho_G$ of 0.01 (Plesset and Prosperetti 1977) coupled with the value of D given above lead to complete solution of a $10\mu m$ bubble in about

2.5s. Though short, this is a long time by the standards of most bubble dynamic phenomena.

The fact that a microbubble should dissolve within seconds leaves unresolved the question of why cavitation nuclei persist indefinitely. One possible explanation is that the interface is immobilized by the effects of surface contamination. Another is that the bubble is imbedded in a solid particle in a way that inhibits the solution of the gas, the so-called Harvey nucleus. These issues were discussed previously in Section 1.12.

Finally we note that there is an important mass diffusion effect caused by ambient pressure oscillations in which nonlinearities can lead to bubble growth even in a subsaturated liquid. This is known as "rectified diffusion" and is discussed later in Section 4.9.

2.7 Thermal Effects on Growth

In Sections 2.4 through 2.6 some of the characteristics of bubble dynamics in the absence of thermal effects were explored. It is now necessary to examine the regime of validity of these analyses, and it is convenient to first evaluate the magnitude of the thermal term (2.24) which was neglected in Equation (2.12) in order to produce Equation (2.27).

First examine the case of bubble growth. The asymptotic growth rate given by Equation (2.33) is constant and hence in the characteristic case of a constant p_∞, terms (1), (3), (4), (5), and (6) in Equation (2.12) are all either constant or diminishing in magnitude as time progresses. Furthermore, a constant, asymptotic growth rate corresponds to the case

$$n = 1; \quad R^* = \{2(p_V - p_\infty^*)/3\rho_L\}^{\frac{1}{2}} \tag{2.59}$$

in Equation (2.21). Consequently, according to Equation (2.24), the thermal term (2) in its linearized form for small $(T_\infty - T_B)$ will be given by

$$\text{term}(2) = \Sigma(T_\infty)C(1)R^* t^{\frac{1}{2}} \tag{2.60}$$

Under these conditions, even if the thermal term is initially negligible, it will gain in magnitude relative to all the other terms and will ultimately affect the growth in a major way. Parenthetically it should be added that the Plesset-Zwick assumption of a small thermal boundary layer thickness, δ_T, relative to R can be shown to hold throughout the inertially controlled growth period since δ_T increases like $(\alpha_L t)^{\frac{1}{2}}$ whereas R is increasing linearly with t. Only under circumstances of very slow growth might the assumption be violated.

Using the relation (2.60), one can define a critical time, t_{c1} (called the first critical time), during the growth when the order of magnitude of term (2) becomes equal to the order of magnitude of the retained terms, as represented by $(\dot{R})^2$. This first critical time is given by

$$t_{c1} = \frac{(p_V - p_\infty^*)}{\rho_L} \cdot \frac{1}{\Sigma^2} \tag{2.61}$$

where the constants of order unity have been omitted for clarity. Thus t_{c1} depends not only on the tension $(p_V - p_\infty^*)/\rho_L$ but also on $\Sigma(T_\infty)$, a purely thermophysical quantity that is a function only of the liquid temperature. Recalling Equation (2.25),

$$\Sigma(T) = \frac{L^2 \rho_V^2}{\rho_L^2 c_{PL} T_\infty \alpha_L^{\frac{1}{2}}} \tag{2.62}$$

it can be anticipated that Σ^2 will change by many, many orders of magnitude in a given liquid as the temperature T_∞ is varied from the triple point to the critical point since Σ^2 is proportional to $(\rho_V/\rho_L)^4$. As a result the critical time, t_{c1}, will vary by many orders of magnitude. Some values of Σ for a number of liquids are plotted in Figure 2.6 as a function of the reduced temperature T/T_C and in Figure 2.7 as a function of the vapor pressure. As an example, consider a typical cavitating flow experiment in a water tunnel with a tension of the order of 10^4 kg/m s^2. Since water at $20°C$ has a value of Σ of about 1 $m/s^{\frac{3}{2}}$, the first critical time is of the order of $10s$, which is very much longer than the time of growth of bubbles. Hence the

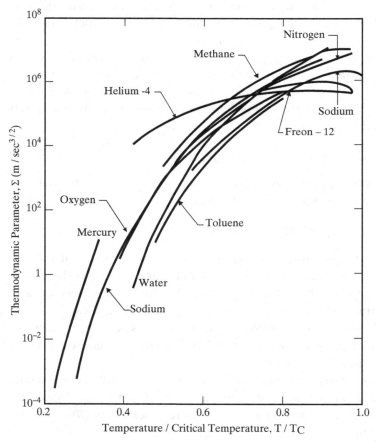

Figure 2.6 Values of the thermodynamic parameter, Σ, for various saturated liquids as a function of the reduced temperature, T/T_C.

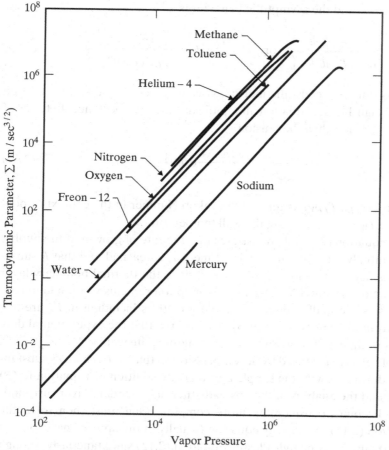

Figure 2.7 Values of the thermodynamic parameter, Σ, for various saturated liquids as a function of the vapor pressure (in $kg/m\ s^2$).

bubble growth occurring in this case is unhindered by thermal effects; it is "inertially controlled" growth. If, on the other hand, the tunnel water were heated to $100°C$ or, equivalently, one observed bubble growth in a pot of boiling water at superheat of $2°K$, then since $\Sigma \approx 10^3\ m/s^{\frac{3}{2}}$ at $100°C$ the first critical time would be $10\mu s$. Thus virtually all the bubble growth observed would be "thermally controlled."

2.8 Thermally Controlled Growth

When the first critical time is exceeded it is clear that the relative importance of the various terms in the Rayleigh-Plesset Equation, (2.12), will change. The most important terms become the driving term (1) and the thermal term (2) whose magnitude is much larger than that of the inertial terms (4). Hence if the tension $(p_V - p_\infty^*)$ remains constant, then the solution using the form of Equation (2.24) for the thermal term must have $n = \frac{1}{2}$ and the asymptotic behavior is

$$R = \frac{(p_V - p_\infty^*)t^{\frac{1}{2}}}{\rho_L \Sigma(T_\infty)C\left(\frac{1}{2}\right)} \quad \underline{\text{or}} \quad n = \frac{1}{2}; \quad R^* = \frac{(p_V - p_\infty^*)}{\rho_L \Sigma(T_\infty)C\left(\frac{1}{2}\right)} \tag{2.63}$$

Consequently, as time proceeds, the inertial, viscous, gaseous, and surface tension terms in the Rayleigh-Plesset equation all rapidly decline in importance. In terms of the superheat, ΔT, rather than the tension

$$R = \frac{1}{2C\left(\frac{1}{2}\right)} \frac{\rho_L c_{PL} \Delta T}{\rho_V L} (\alpha_L t)^{\frac{1}{2}} \tag{2.64}$$

where the group $\rho_L c_{PL} \Delta T / \rho_V L$ is termed the Jakob Number in the context of pool boiling and $\Delta T = T_W - T_\infty$, T_W being the wall temperature.

The result, Equation (2.63), demonstrates that the rate of growth of the bubble decreases substantially after the first critical time, t_{c1}, is reached and that R subsequently increases like $t^{\frac{1}{2}}$ instead of t. Moreover, since the thermal boundary layer also increases like $(\alpha_L t)^{\frac{1}{2}}$, the Plesset-Zwick assumption remains valid indefinitely. An example of this thermally inhibited bubble growth is including in Figure 2.8, which is taken from Dergarabedian (1953). We observe that the experimental data and calculations using the Plesset-Zwick method agree quite well.

When bubble growth is caused by decompression so that $p_\infty(t)$ changes substantially with time during growth, the simple approximate solution of Equation (2.63) no longer holds and the analysis of the unsteady thermal boundary layer surrounding the bubble becomes considerably more complex. One must then solve the diffusion Equation (2.15), the energy equation (usually in the approximate form of Equation (2.17)) and the Rayleigh-Plesset Equation (2.12) simultaneously, though

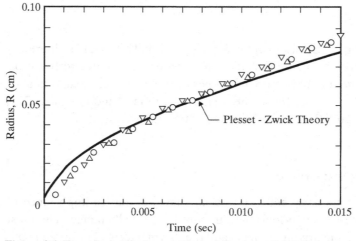

Figure 2.8 Experimental observations of the growth of three vapor bubbles ($\bigcirc$, $\triangle$, $\triangledown$) in superheated water at $103.1°C$ compared with the growth expected using the Plesset-Zwick theory (adapted from Dergarabedian 1953).

for the thermally controlled growth being considered here, most of the terms in Equation (2.12) become negligible so that the simplification, $p_V(T_B) = p_\infty(t)$, is usually justified. When p_∞ is a constant this reduces to the problem treated by Plesset and Zwick (1952) and later addressed by Forster and Zuber (1954) and Scriven (1959). Several different approximate solutions to the general problem of thermally controlled bubble growth during liquid decompression have been put forward by Theofanous et al. (1969), Jones and Zuber (1978) and Cha and Henry (1981). Theofanous et al. include nonequilibrium thermodynamic effects on which we comment in the following section. If these are ignored, then all three analyses yield qualitatively similar results which also agree quite well with the experimental data of Hewitt and Parker (1968) for bubble growth in liquid nitrogen. Figure 2.9 presents a typical example of the data of Hewitt and Parker and a comparison with the three analytical treatments mentioned above.

Several other factors can complicate and alter the dynamics of thermally controlled growth, and these are discussed in the sections which follow. Nonequilibrium effects are addressed in Section 2.9. More important are the modifications to the heat transfer mechanisms at the bubble surface that can be caused by surface

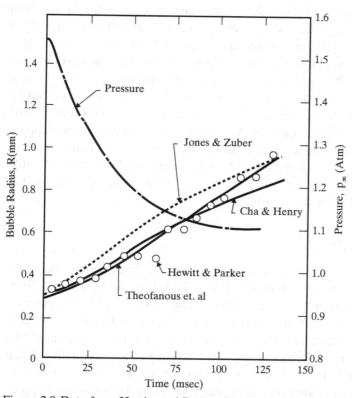

Figure 2.9 Data from Hewitt and Parker (1968) on the growth of a vapor bubble in liquid nitrogen (pressure/time history also shown) and comparison with the analytical treatments by Theofanous et al. (1969), Jones and Zuber (1978), and Cha and Henry (1981).

instabilities or by convective heat transfer. These are reviewed in Sections 2.10 and 2.12.

2.9 Nonequilibrium Effects

One factor that could affect the dynamics of thermally controlled growth is whether or not the liquid at the interface is in thermal equilibrium with the vapor in the bubble. Most of the analyses assume that the temperature of the liquid at the interface, T_{LS}, is the temperature of the saturated vapor in the bubble, T_B. Theofanous et al. (1969) have suggested that this might not be the case because of the high evaporation rate. They employ an accommodation coefficient, Λ, defined (Schrage 1953) by

$$G_V = \Lambda \left[\left(\frac{1}{2\pi K_V} \right)^{\frac{1}{2}} \left(\frac{p_V(T_{LS})}{T_{LS}^{\frac{1}{2}}} - \frac{p_V(T_B)}{T_B^{\frac{1}{2}}} \right) \right] \tag{2.65}$$

where G_V is the evaporative mass flux and K_V is the gas constant of the vapor. For a chosen value of Λ this effectively defines a temperature discontinuity at the interface. Clearly $\Lambda = \infty$ corresponds to the previously assumed equilibrium condition. Plesset and Prosperetti (1977) demonstrate that if Λ is of order unity then the nonequilibrium correction is of the order of the Mach number of the bubble wall motion and is therefore negligible except, perhaps, near the end of a violent bubble collapse (see Fujikawa and Akamatsu 1980 and Section 3.2). On the other hand, if Λ is much smaller than unity, significant nonequilibrium effects might be encountered.

Theofanous et al. (1969) explore the effects of small values of Λ theoretically. They confirm that values of order unity do not yield bubble histories that differ by very much from those that assume equilibrium. Values of Λ of the order of 0.01 did produce substantial differences. However, the results using equilibrium appear to compare favorably with the experimental results as shown in Figure 2.9. This suggests that nonequilibrium effects have little effect on thermally controlled bubble growth though the issue is not entirely settled since some studies do suggest that values of Λ as low as 0.01 may be possible.

2.10 Convective Effects

Another way in which the rate of heat transfer to the interface may be changed is by convection caused by relative motion between the bubble and the liquid. Such enhancement of the heat transfer rate is normally represented by a Nusselt number, Nu, defined as the ratio of the actual heat transfer rate divided by the rate of heat transfer by conduction. Therefore in the present context the factor, Nu, should be included as a multiplier in the thermal term of the Rayleigh-Plesset equation. Then one seeks a relationship between Nu and the Peclet number, $Pe = WR/\alpha_L$, where W is the typical translational velocity of the bubble relative to the liquid. The appropriate relationship for a growing and translating bubble is not known; analytically this represents a problem that is substantially more complex than that tackled by

Plesset and Zwick. Nevertheless, it is of interest to speculate on the form of $Nu(Pe)$ and observe the consequences for the bubble growth rate. Therefore let us assume that this relationship takes the approximate form common in many convective heat transfer problems:

$$Nu = 1 \qquad \text{for } Pe \ll 1$$
$$= C Pe^m \quad \text{for } Pe \gg 1 \tag{2.66}$$

where C is some constant of order unity. We must also decide on the form of the relative velocity, W, which could have several causes. In either a cavitating flow or in pool boiling it could be due to pressure gradients within the liquid due to acceleration of the liquid. It could also be caused by the presence of nearby solid boundaries.

Despite the difficulties of accurate assessment of the convective heat transfer effects, let us consider the qualitative effects of two possible translational motions on a bubble growing like $R = R^* t^n$. The first effect is that due to buoyancy; the relative velocity, W, caused by buoyancy in the absence of viscous drag will be given by $W \propto gt$ with a factor of proportionality of order one. The viscous drag on the bubble will have little effect so long as $\nu_L t \ll R^2$. The second example is a bubble growing on a solid wall where the effective convective velocity is roughly given by dR/dt and hence $W \propto R^* t^{n-1}$. Thus the Peclet numbers for the two cases are respectively

$$\frac{R^* g t^{n+1}}{\alpha_L} \quad \text{and} \quad \frac{(R^*)^2 t^{2n-1}}{\alpha_L} \tag{2.67}$$

Consider first the case of inertially controlled growth for which $n = 1$. Then it follows that convective heat transfer effects will only occur for $Pe \geq 1$ or for times $t > t_{c2}$ where

$$t_{c2} = \left[\frac{\rho_L \alpha_L^2}{(p_V - p_\infty^*) g^2} \right]^{\frac{1}{4}} \quad \text{and} \quad \frac{\alpha_L \rho_L}{(p_V - p_\infty^*)} \tag{2.68}$$

respectively where the asymptotic growth rate given by Equation (2.33) has been used. Consequently, the convective enhancement of the heat transfer will only occur during the inertially controlled growth if $t_{c2} < t_{c1}$ and this requires that

$$p_V - p_\infty^* > \rho_L \left[\frac{\Sigma^4 \alpha_L}{g} \right]^{\frac{2}{5}} \quad \text{and} \quad \rho_L (\Sigma^2 \alpha_L)^{\frac{1}{2}} \tag{2.69}$$

respectively. Since Σ increases rapidly with temperature it is *much* more likely that these inequalities will be true at low reduced temperatures than at high reduced temperatures. For example, in water at $20°C$ the right-hand sides of Inequality (2.69) are respectively 30 and 4 $kg/msec^2$, very small tensions (and correspondingly minute superheats) that could readily occur. If the tension is larger than this critical value, then convective effects would become important. On the other hand, in water at $100°C$ the values are respectively equivalent to superheats of $160°K$ and $0.5°K$, which are less likely to occur.

It follows that in each of the two bubble motions assumed there is some temperature below which one would expect Pe to reach unity prior to t_{c1}. The question is what happens thereafter, for clearly the thermal effect that would otherwise begin at t_{c1} is now going to be altered by the enhanced heat transfer. When $Pe > 1$ the thermal term in the Rayleigh-Plesset equation will no longer grow like $t^{\frac{1}{2}}$ but will increase like $t^{\frac{1}{2}}/Nu$ which, according to the relations (2.67), is like $t^{\frac{1}{2}-2m}$ and $t^{\frac{1}{2}-m}$ for the two bubble motions. If, as in many convective heat transfer problems, $m = \frac{1}{2}$, it would follow that thermal inhibition of the growth would be eliminated and the inertially controlled growth would continue indefinitely.

Finally, consider the other possible scenario in which convective heat transfer effects might influence the thermally controlled growth in the event that $t_{c2} > t_{c1}$. Given $n = \frac{1}{2}$, the Peclet number for buoyancy-induced motion would become unity at

$$t_{c3} = \left[\frac{\alpha_L \rho_L \Sigma}{g(p_V - p_\infty^*)} \right]^{2/3} \tag{2.70}$$

using Equation (2.63). Consequently, convective heat transfer could alter the form of the thermally controlled growth after $t = t_{c3}$; indeed, it is possible that inertially controlled growth could resume after t_{c3} if $m > \frac{1}{4}$. In the other example of bubble growth at a wall, the Peclet number would remain at the value less than unity which it had attained at t_{c1}. Consequently, the convective heat transfer effects would delay the onset of thermally inhibited growth indefinitely if $(p_V - p_\infty^*) \gg \rho_L(\Sigma^2 \alpha_L)^{\frac{1}{2}}$ but would have little or no effect on either the onset or form of the thermally controlled growth if the reverse were true.

2.11 Surface Roughening Effects

Another important phenomenon that can affect the heat transfer process at the interface during bubble growth (and therefore affect the bubble growth rate) is the development of an instability on the interface. If the bubble surface becomes rough and turbulent, the increase in the effective surface area and the unsteady motions of the liquid near that surface can lead to a substantial enhancement of the rate of heat transfer to the interface. The effect is to delay (perhaps even indefinitely) the point at which the rate of growth is altered by thermal effects. This is one possible explanation for the phenomenon of vapor explosions which are essentially the result of an extended period of inertially controlled bubble growth.

Shepherd and Sturtevant (1982) and Frost and Sturtevant (1986) have examined rapidly growing nucleation bubbles near the limit of superheat and have found growth rates substantially larger than expected when the bubble was in the thermally controlled growth phase. The experiments examined bubble growth within droplets of superheated liquid suspended in another immiscible liquid. Typical photographs are shown in Figure 2.10 and reveal that the surfaces of the bubbles are rough and irregular. The enhancement of the heat transfer caused by this roughening is probably responsible for the larger growth rates. Shepherd and Sturtevant

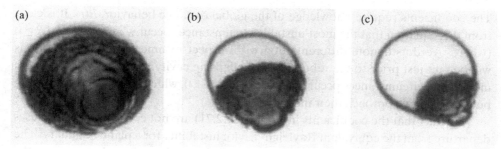

Figure 2.10 Typical photographs of a rapidly growing bubble in a droplet of superheated ether suspended in glycerine. The bubble is the dark, rough mass; the droplet is clear and transparent. The photographs, which are of different events, were taken 31, 44, and 58 μs after nucleation and the droplets are approximately $2mm$ in diameter. Reproduced from Frost and Sturtevant (1986) with the permission of the authors.

(1982) attribute the roughness to the development of a baroclinic interfacial instability similar to the Landau-Darrieus instability of flame fronts. It is also of interest to note that Frost and Sturtevant report that the instability could be suppressed by increasing the ambient pressure and therefore the temperature and density within the bubble. At an ambient pressure of 2 *bar*, the onset of the instability could be observed on the surface of ether bubbles and was accompanied by a jump in the radiated pressure associated with the sudden acceleration in the growth rate. At higher ambient pressures the instability could be completely suppressed. This occurs because the growth rate of the instability increases with the rate of growth of the bubble, and both are significantly reduced at the higher ambient pressures. It may be that, under other circumstances, the Rayleigh-Taylor instabilities described in Section 2.12 could give rise to a similar effect.

2.12 Nonspherical Perturbations

Apart from the phenomena described in the preceding section, it has, thus far, been tacitly assumed that the bubble remains spherical during the growth or collapse process; in other words, it has been assumed that the bubble is *stable* to nonspherical distortions. There are, however, circumstances in which this is not true, and the subsequent departure from a smooth spherical shape can have important practical consequences.

The stability to nonspherical disturbances has been investigated from a purely hydrodynamic point of view by Birkhoff (1954) and Plesset and Mitchell (1956), among others. These analyses essentially examine the spherical equivalent of the Rayleigh-Taylor instability; they do not include thermal effects. If the inertia of the gas in the bubble is assumed to be negligible, then the amplitude, $a(t)$, of a spherical harmonic distortion of order n $(n > 1)$ will be governed by the equation:

$$\frac{d^2a}{dt^2} + \frac{3}{R}\frac{dR}{dt}\frac{da}{dt} - \left[\frac{(n-1)}{R}\frac{d^2R}{dt^2} - (n-1)(n+1)(n+2)\frac{S}{\rho_L R^3}\right] a = 0 \qquad (2.71)$$

The coefficients require knowledge of the global dynamic behavior, $R(t)$. It is clear from this equation that the most unstable circumstances occur when $\dot{R} < 0$ and $\ddot{R} \geq 0$ (where overdots denote differentiation with respect to time, t). These conditions will be met just prior to the rebound of a collapsing cavity. On the other hand, the most stable circumstances occur when $\dot{R} > 0$ and $\ddot{R} < 0$, which is the case for growing bubbles as they approach their maximum size.

The fact that the coefficients in Equation (2.71) are not constant in time causes departure from the equivalent Rayleigh-Taylor instability for a plane boundary. The coefficient of a is not greatly dissimilar from the case of the plane boundary in the sense that instability is promoted when $\ddot{R} > 0$ and surface tension has a stabilizing effect. The primary difference is caused by the da/dt term, which can be interpreted as a geometric effect. As the bubble grows the wavelength on the surface increases, and hence the growth of the wave amplitude is lessened. The reverse occurs during collapse.

Plesset and Mitchell (1956) examined the particular case of a vapor/gas bubble initially in equilibrium that is subjected to a step function change in the pressure at infinity. Thermal and viscous effects are assumed to be negligible. The effect of a fixed mass of gas in the bubble will be included in this presentation though it was omitted by Plesset and Mitchell. Note that this simple growth problem for a spherical bubble was solved for $R(t)$ in Section 2.4. One feature of that solution that is important in this context is that $\ddot{R} \geq 0$. It is this feature that gives rise to the instability. However, in any real scenario, the initial acceleration phase for which $\ddot{R} \geq 0$ is of limited duration, so the issue will be whether or not the instability has sufficient time during the acceleration phase for significant growth to occur.

It transpires that it is more convenient to rewrite Equation (2.71) using $y = R/R_o$ as the independent variable rather than t. Then $a(y)$ must satisfy

$$A(y)\frac{d^2 a}{dy^2} + \frac{B(y)}{y}\frac{da}{dy} - \frac{(n-1)C(y)a}{y^3} = 0 \tag{2.72}$$

where

$$A(y) = \frac{2}{3}\left(1 - \frac{1}{y^3}\right) - \frac{\beta_1}{y}\left(1 - \frac{1}{y^2}\right) + \frac{\beta_2}{y^3}\ln y \tag{2.73}$$

$$B(y) = 2 - \frac{1}{y^3} - \frac{\beta_1}{y}\left(\frac{5}{2} - \frac{3}{2y^2}\right) + \frac{\beta_2}{y^3}\left(1 + \frac{3}{2}\ln y\right) \tag{2.74}$$

$$C(y) = \frac{1}{y^2} - \frac{3\beta_1}{2y^2} + \frac{\beta_2}{y^2}\left(1 - \frac{3}{2}\ln y\right) + \frac{\beta_1}{2}\{1 - (n+1)(n+2)\} \tag{2.75}$$

and the parameters

$$\beta_1 = \frac{2S}{R_o(p_V - p_\infty)}, \quad \beta_2 = \frac{p_{Go}}{(p_V - p_\infty)} \tag{2.76}$$

represent the effects of surface tension and gas content respectively. Note that a positive value of

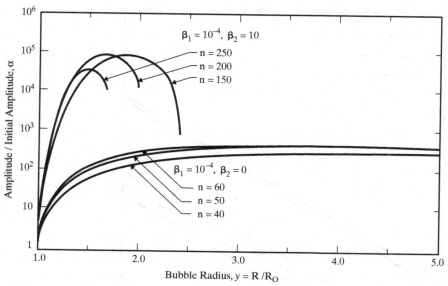

Figure 2.11 Examples of the growth of the amplitude, a, of a spherically harmonic disturbance (of order n as indicated) on the surface of a growing cavitation bubble for two typical choices of the surface tension and gas content parameters, β_1 and β_2.

$$(p_V - p_\infty)(1 - \beta_1 + \beta_2) \qquad (2.77)$$

implies bubble growth following $t = 0$ whereas a negative value implies collapse.

Some typical numerical integrations of Equation (2.72) in cases of bubble growth are shown in Figure 2.11 where the amplitude scale is arbitrary. Plesset and Mitchell performed hand calculations for small n and found only minor amplification during growth. However, as can be anticipated from Equation (2.72), the amplitudes may be much larger for large n. It can be seen from Figure 2.11 that the amplitude of the disturbance reaches a peak and then decays during growth. For given values of the parameters β_1 and β_2, there exists a particular spherical harmonic, $n = n_A$, which achieves the maximum amplitude, a; in Figure 2.11 we chose to display data for n values which bracket n_A. The dependence of n_A on β_1 and β_2 is shown in Figure 2.12. A slightly different value of n denoted by n_B gives the maximum value of a/y. Since the latter quantity rather than a represents a measure of the wave amplitude to wavelength ratio, Figure 2.12 also shows the dependence of n_B on β_1 and β_2. To complete the picture, Figure 2.13 presents values of $(a)_{max}$, $(a/y)_{max}$ and the sizes of the bubble $(y)_{a=max}$ and $(y)_{a/y=max}$ at which these maxima occur.

In summary, it can be seen that the initial acceleration phase of bubble growth in which $\ddot{R} \geq 0$ is unstable to spherical harmonic perturbations of fairly high order, n. On the other hand, visual inspection of Equation (2.71) is sufficient to conclude that the remainder of the growth phase during which $\dot{R} > 0$, $\ddot{R} < 0$ is stable to all spherical harmonic perturbations. So, if inadequate time is available for growth of the perturbations during the acceleration phase, then the bubble will remain unperturbed throughout its growth. In their experiments on underwater explosions, Reynolds and Berthoud (1981) observed bubble surface instabilities during the acceleration

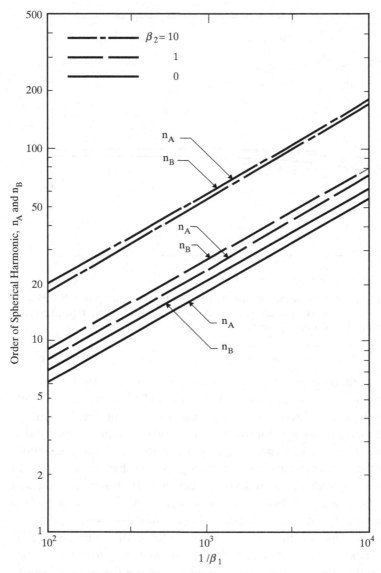

Figure 2.12 The orders of the spherical harmonic disturbances that, during bubble growth, produce (i) the maximum disturbance amplitude (n_A) and (ii) the maximum ratio of disturbance amplitude to bubble radius (n_B) for various surface tension and gas content parameters, β_1 and β_2.

phase that did correspond to fairly large n of the order of 10. They also evaluate the duration of the acceleration phase in their experiments and demonstrate, using an estimated growth rate, that this phase is long enough for significant roughening of the surface to occur. However, their bubbles become smooth again in the second, deceleration phase of growth. The bubbles examined by Reynolds and Berthoud were fairly large, 2.5 cm to 4.5 cm in radius. A similar acceleration phase instability

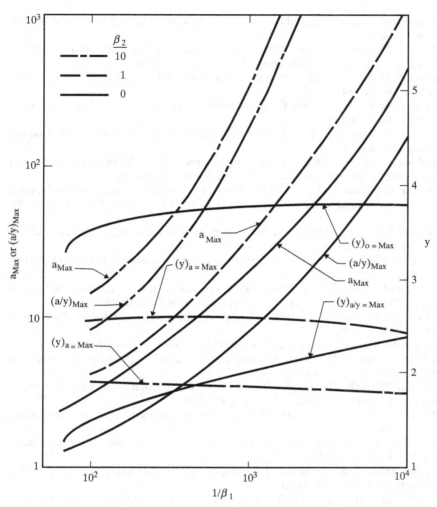

Figure 2.13 The maximum amplification, a_{max}, and the maximum ratio of amplitude to bubble radius, $(a/y)_{max}$, for spherical harmonic disturbances on the surface of a growing bubble for various surface tension and gas content parameters, β_1 and β_2. Also shown are the bubble sizes, $(y)_{a=max}$ and $(y)_{a/y=max}$, at which these maxima occur.

has not, to the author's knowledge, been reported for the smaller bubbles typical of most cavitation experiments. This could either be the result of a briefer acceleration phase or the greater stabilizing effect of surface tension in smaller bubbles.

References

Birkhoff, G. (1954). Note on Taylor instability. *Quart. Appl. Math.*, **12**, 306–309.

Blake, F.G. (1949). The onset of cavitation in liquids: I. *Acoustics Res. Lab., Harvard Univ., Tech. Memo. No. 12*.

Ceccio, S.L. and Brennen, C.E. (1991). Observations of the dynamics and acoustics of travelling bubble cavitation. *J. Fluid Mech.*, **233**, 633–660.

Cha, Y.S. and Henry, R.E. (1981). Bubble growth during decompression of a liquid. *ASME J. Heat Transfer*, **103**, 56–60.

Daily, J.W. and Johnson, V.E., Jr. (1956). Turbulence and boundary layer effects on cavitation inception from gas nuclei. *Trans. ASME*, **78**, 1695–1706.

Dergarabedian, P. (1953). The rate of growth of vapor bubbles in superheated water. *ASME J. Appl. Mech.*, **20**, 537–545.

Epstein, P.S. and Plesset, M.S. (1950). On the stability of gas bubbles in liquid-gas solutions. *J. Chem. Phys.*, **18**, 1505–1509.

Forster, H.K. and Zuber, N. (1954). Growth of a vapor bubble in a superheated liquid. *J. Appl. Phys.*, **25**, No.4, 474–478.

Frost, D. and Sturtevant, B. (1986). Effects of ambient pressure on the instability of a liquid boiling explosively at the superheat limit. *ASME J. Heat Transfer*, **108**, 418–424.

Fujikawa, S. and Akamatsu, T. (1980). Effects of the non-equilibrium condensation of vapour on the pressure wave produced by the collapse of a bubble in a liquid. *J. Fluid Mech.*, **97**, 481–512.

Hewitt, H.C. and Parker, J.D. (1968). Bubble growth and collapse in liquid nitrogen. *ASME J. Heat Transfer*, **90**, 22–26.

Jones, O.C. and Zuber, N. (1978). Bubble growth in variable pressure fields. *ASME J. Heat Transfer*, **100**, 453–459.

Neppiras, E.A. and Noltingk, B.E. (1951). Cavitation produced by ultrasonics: theoretical conditions for the onset of cavitation. *Proc. Phys. Soc., London*, **64B**, 1032–1038.

Noltingk, B.E. and Neppiras, E.A. (1950). Cavitation produced by ultrasonics. *Proc. Phys. Soc., London*, **63B**, 674–685.

Plesset, M.S. (1949). The dynamics of cavitation bubbles. *ASME J. Appl. Mech.*, **16**, 228–231.

Plesset, M.S. and Zwick, S.A. (1952). A nonsteady heat diffusion problem with spherical symmetry. *J. Appl. Phys.*, **23**, No. 1, 95–98.

Plesset, M.S. and Mitchell, T.P. (1956). On the stability of the spherical shape of a vapor cavity in a liquid. *Quart. Appl. Math.*, **13**, No. 4, 419–430.

Plesset, M.S. and Prosperetti, A. (1977). Bubble dynamics and cavitation. *Ann. Rev. Fluid Mech.*, **9**, 145–185.

Poritsky, H. (1952). The collapse or growth of a spherical bubble or cavity in a viscous fluid. *Proc. First Nat. Cong. in Appl. Math.*, 813–821.

Rayleigh, Lord. (1917). On the pressure developed in a liquid during the collapse of a spherical cavity. *Phil. Mag.*, **34**, 94–98.

Reynolds, A.B. and Berthoud, G. (1981). Analysis of EXCOBULLE two-phase expansion tests. *Nucl. Eng. and Design*, **67**, 83–100.

Schrage, R.W. (1953). *A theoretical study of interphase mass transfer*. Columbia Univ. Press.

Scriven, L.E. (1959). On the dynamics of phase growth. *Chem. Eng. Sci.*, **10**, 1–13.

Shepherd, J.E. and Sturtevant, B. (1982). Rapid evaporation near the superheat limit. *J. Fluid Mech.*, **121**, 379–402.

Theofanous, T., Biasi, L., Isbin, H.S., and Fauske, H. (1969). A theoretical study on bubble growth in constant and time-dependent pressure fields. *Chem. Eng. Sci.*, **24**, 885–897.

3 Cavitation Bubble Collapse

3.1 Introduction

In the preceding chapter some of the equations of bubble dynamics were developed and applied to problems of bubble growth. In this chapter we continue the discussion of bubble dynamics but switch attention to the dynamics of collapse and, in particular, consider the consequences of the violent collapse of vapor-filled cavitation bubbles.

3.2 Bubble Collapse

Bubble collapse is a particularly important subject because of the noise and material damage that can be caused by the high velocities, pressures, and temperatures that may result from that collapse. The analysis of Section 2.4 allowed approximate evaluation of the magnitudes of those velocities, pressures, and temperatures (Equations (2.36), (2.38), (2.39)) under a number of assumptions including that the bubble remains spherical. It will be shown in Section 3.5 that collapsing bubbles do not remain spherical. Moreover, as we shall see in Chapter 7, bubbles that occur in a cavitating flow are often far from spherical. However, it is often argued that the spherical analysis represents the maximum possible consequences of bubble collapse in terms of the pressure, temperature, noise, or damage potential. Departure from sphericity can diffuse the focus of the collapse and reduce the maximum pressures and temperatures that might result.

When a cavitation bubble grows from a small nucleus to many times its original size, the collapse will begin at a maximum radius, R_M, with a partial pressure of gas, p_{GM}, which is very small indeed. In a typical cavitating flow R_M is of the order of 100 times the original nuclei size, R_o. Consequently, if the original partial pressure of gas in the nucleus was about 1 bar the value of p_{GM} at the start of collapse would be about 10^{-6} bar. If the typical pressure depression in the flow yields a value for $(p^*_\infty - p_\infty(0))$ of, say, 0.1 bar it would follow from Equation (2.38) that the maximum pressure generated would be about 10^{10} bar and the maximum temperature would be 4×10^4 times the ambient temperature! Many factors, including the diffusion of gas from the liquid into the bubble and the effect of liquid compressibility, mitigate this result. Nevertheless, the calculation illustrates the potential for the generation of high pressures and temperatures during collapse and the potential for the generation of shock waves and noise.

Early work on collapse focused on the inclusion of liquid compressibility in order to learn more about the production of shock waves. Herring (1941) introduced the first-order correction for liquid compressibility assuming the Mach number of collapse motion, $|dR/dt|/c$, was much less than unity and neglecting any noncondensable gas or thermal effects so that the pressure in the bubble remains constant. Later, Schneider (1949) treated the same, highly idealized problem by numerically solving the equations of compressible flow up to the point where the Mach number of collapse, $|dR/dt|/c$, was about 2.2. Gilmore (1952) (see also Trilling 1952) showed that one could use the approximation introduced by Kirkwood and Bethe (1942) to obtain analytic solutions that agreed with Schneider's numerical results up to that Mach number. Parenthetically we note that the Kirkwood-Bethe approximation assumes that wave propagation in the liquid occurs at sonic speed, c, relative to the liquid velocity, u, or, in other words, at $c + u$ in the absolute frame (see also Flynn 1966). Figure 3.1 presents some of the results obtained by Herring (1941), Schneider (1949), and Gilmore (1952). It demonstrates how, in the idealized problem, the Mach number of the bubble surface increases as the bubble radius decreases. The line marked "incompressible" corresponds to the case in which the compressibility of the liquid has been neglected in the equation of motion (see Equation (2.36)). The slope is approximately $-3/2$ since $|dR/dt| \propto R^{-\frac{3}{2}}$. Note that compressibility tends to lessen the velocity of collapse. We note that Benjamin

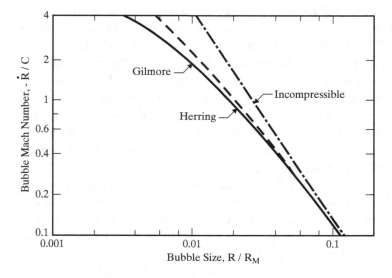

Figure 3.1 The bubble surface Mach number, $-\dot{R}/c$, plotted against the bubble radius (relative to the initial radius) for a pressure difference, $p_\infty - p_{GM}$, of $0.517\ bar$. Results are shown for the incompressible analysis and for the methods of Herring (1941) and Gilmore (1952). Schneider's (1949) numerical results closely follow Gilmore's curve up to a Mach number of 2.2.

(1958) also investigated analytical solutions to this problem at higher Mach numbers for which the Kirkwood-Bethe approximation becomes quite inaccurate.

When the bubble contains some noncondensable gas or when thermal effects become important, the solution becomes more complex since the pressure in the bubble is no longer constant. Under these circumstances it would clearly be very useful to find some way of incorporating the effects of liquid compressibility in a modified version of the Rayleigh-Plesset equation. Keller and Kolodner (1956) proposed the following modified form in the absence of thermal, viscous, or surface tension effects:

$$\left(1 - \frac{1}{c}\frac{dR}{dt}\right)R\frac{d^2R}{dt^2} + \frac{3}{2}\left(1 - \frac{1}{3c}\frac{dR}{dt}\right)\left(\frac{dR}{dt}\right)^2$$
$$= \left(1 + \frac{1}{c}\frac{dR}{dt}\right)\frac{1}{\rho_L}\left[p_B - p_\infty - p_c(t + R/c)\right] + \frac{R}{\rho_L c}\frac{dp_B}{dt} \tag{3.1}$$

where $p_c(t)$ denotes the variable part of the pressure in the liquid at the location of the bubble center in the absence of the bubble. Other forms have been suggested and the situation has recently been reviewed by Prosperetti and Lezzi (1986), who show that a number of the suggested equations are equally valid in that they are all accurate to the first or linear order in the Mach number, $|dR/dt|/c$. They also demonstrate that such modified Rayleigh-Plesset equations are quite accurate up to Mach numbers of the order of 0.3. At higher Mach numbers the compressible liquid field equations must be solved numerically.

However, as long as there is some gas present to decelerate the collapse, the primary importance of liquid compressibility is not the effect it has on the bubble dynamics (which is slight) but the role it plays in the formation of shock waves during the rebounding phase that follows collapse. Hickling and Plesset (1964) were the first to make use of numerical solutions of the compressible flow equations to explore the formation of pressure waves or shocks during the rebound phase. Figure 3.2 presents an example of their results for the pressure distributions in the liquid before (left) and after (right) the moment of minimum size. The graph on the right clearly shows the propagation of a pressure pulse or shock away from the bubble following the minimum size. As indicated in that figure, Hickling and Plesset concluded that the pressure pulse exhibits approximately geometric attentuation (like r^{-1}) as it propagates away from the bubble. Other numerical calculations have since been carried out by Ivany and Hammitt (1965), Tomita and Shima (1977), and Fujikawa and Akamatsu (1980), among others. Ivany and Hammitt (1965) confirmed that neither surface tension nor viscosity play a significant role in the problem. Effects investigated by others will be discussed in the following section.

These later works are in accord with the findings of Hickling and Plesset (1964) insofar as the development of a pressure pulse or shock is concerned. It appears that, in most cases, the pressure pulse radiated into the liquid has a peak pressure amplitude, p_P, which is given roughly by

$$p_P \approx 100 R_M p_\infty / r \tag{3.2}$$

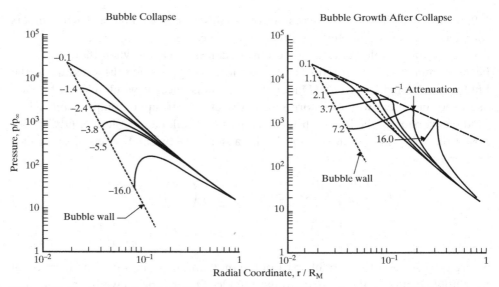

Figure 3.2 Typical results of Hickling and Plesset (1964) for the pressure distributions in the liquid before collapse (left) and after collapse (right) (without viscosity or surface tension). The parameters are $p_\infty = 1$ *bar*, $\gamma = 1.4$, and the initial pressure in the bubble was 10^{-3} *bar*. The values attached to each curve are proportional to the time before or after the minimum size.

Though Akulichev (1971) found much stronger attentuation in the far field, it seems clear that Equation (3.2) gives the order of magnitude of the strong pressure pulse, which might impinge on a solid surface a few radii away. For example, if $p_\infty \approx 1$ *bar* this implies a substantial pulse of 100 *bar* at a distance of one maximum bubble radius away (at $r = R_M$). Experimentally, Fujikawa and Akamatsu (1980) found shock intensities at the wall of about 100 *bar* when the collapsing bubble was about a maximum radius away from the wall. We note that much higher pressures are momentarily experienced in the gas of the bubble, but we shall delay discussion of this feature of the results until later.

All of these analyses assume spherical symmetry. Later we will focus attention on the stability of shape of a collapsing bubble before continuing discussion of the origins of cavitation damage.

3.3 Thermally Controlled Collapse

Before examining thermal effects during the last stages of collapse, it is important to recognize that bubbles could experience thermal effects early in the collapse in the same way as was discussed for growing bubbles in Section 2.7. As one can anticipate, this would negate much of the discussion in the preceding and following sections since if thermal effects became important early in the collapse phase, then the subsequent bubble dynamics would be of the benign, thermally controlled type.

Consider a bubble of radius, R_o, initially at rest at time, $t = 0$, in liquid at a pressure, p_∞. Collapse is initiated by increasing the ambient liquid pressure to p_∞^*. From the Rayleigh-Plesset equation the initial motion in the absence of thermal effects has the form

$$R/R_o = 1 - p_c t^2 + 0(t^3) \tag{3.3}$$

where p_c is the collapse motivation defined as

$$p_c = \left[\frac{(p_\infty^* - pv)}{2\rho_L R_o^2} + \frac{S}{\rho_L R_o^3} - \frac{3m_G T_\infty K_G}{8\pi \rho_L R_o^5} \right] \tag{3.4}$$

If this is substituted into the Plesset-Zwick Equation (2.20) to evaluate the thermal term in the Rayleigh-Plesset equation, one obtains a critical time t_{c4}, necessary for development of significant thermal effects given by

$$t_{c4} = (R_o/\Sigma)^{2/3} \tag{3.5}$$

One problem with such an approach is that the Plesset-Zwick assumption of a thermal boundary layer that is thin compared to R will be increasingly in danger of being violated as the boundary layer thickens while the radius decreases. Nevertheless, proceeding with the analysis, it follows that if $t_{c4} \ll t_{TC}$ where t_{TC} is the typical time for collapse (see Section 2.4), then thermally controlled collapse will begin early in the collapse process. It follows that this condition arises if

$$(p_c)^{1/2} \left(\frac{R_o}{\Sigma} \right)^{2/3} \ll 1 \tag{3.6}$$

If this is the case then the initial motion will be effectively dominated by the thermal term and will be of the form

$$R = R_o - \frac{p_c R_o^2}{\Sigma} \left[\frac{4\pi^{1/2}}{\int_0^1 \frac{dy}{(y(1-y))^{1/2}}} \right] t^{1/2} \tag{3.7}$$

where the term in the square bracket is a simple constant of order unity. If Inequality (3.6) is violated, then thermal effects will not begin to become important until later in the collapse process.

3.4 Thermal Effects in Bubble Collapse

Even if thermal effects are negligible for most of the collapse phase, they play a very important role in the final stage of collapse when the bubble contents are highly compressed by the inertia of the inrushing liquid. The pressures and temperatures that are predicted to occur in the gas within the bubble during spherical collapse are very high indeed. Since the elapsed times are so small (of the order of microseconds), it would seem a reasonable approximation to assume that the noncondensable gas in the bubble behaves adiabatically. Typical of the adiabatic calculations is the work of Tomita and Shima (1977), who used the accurate method for handling liquid compressiblity that was first suggested by Benjamin (1958) and obtained maximum gas temperatures as high as $8800°K$ in the bubble center. But, despite the small elapsed times, Hickling (1963) demonstrated that heat transfer between the liquid and the gas is important because of the extremely high temperature gradients and the short distances involved. In later calculations Fujikawa and Akamatsu (1980) included heat transfer and, for a case similar to that of Tomita and Shima, found lower maximum temperatures and pressures of the order of $6700°K$ and 848 *bar* respectively at the bubble center. The gradients of temperature are such that the maximum interface temperature is about $3400°K$. Furthermore, these temperatures and pressures only exist for a fraction of a microsecond; for example, after $2 \ \mu s$ the interface temperature dropped to $300°K$.

Fujikawa and Akamatsu (1980) also explored nonequilibrium condensation effects at the bubble wall which, they argued, could cause additional cushioning of the collapse. They carried out calculations that included an accommodation coefficient similar to that defined in Equation (2.65). As in the case of bubble growth studied by Theofanous et al. (1969), Fujikawa and Akamatsu showed that an accommodation coefficient, Λ, of the order of unity had little effect. Accommodation coefficients of the order of 0.01 were required to observe any significant effect; as we commented in Section 2.9, it is as yet unclear whether such small accommodation coefficients would occur in practice.

Other effects that may be important are the interdiffusion of gas and vapor within the bubble, which could cause a buildup of noncondensable gas at the interface and therefore create a barrier which through the vapor must diffuse in order to condense on the interface. Matsumoto and Watanabe (1989) have examined a similar effect in the context of oscillating bubbles.

3.5 Nonspherical Shape During Collapse

Now consider the collapse of a bubble that contains primarily vapor. As in Section 2.4 we will distinguish between the two important stages of the motion excluding the initial inward acceleration transient. These are

1. the asymptotic form of the collapse in which $\dot{R} \propto R^{-\frac{3}{2}}$, which occurs prior to significant compression of the gas content, and
2. the rebound stage, in which the acceleration, $\ddot{R}$, reverses sign and takes a very large positive value.

The stability characteristics of these two stages are very different. The calculations of Plesset and Mitchell (1956) showed that a bubble in an infinite medium would only be mildly unstable during the first stage in which $\ddot{R}$ is negative; disturbances would only grow at a slow rate due to geometric effects. Note that for small y, Equation (2.72) reduces to

$$\frac{d^2 a}{dy^2} + \frac{3}{2y} \frac{da}{dy} + (n-1)\frac{a}{y^2} = 0 \qquad (3.8)$$

which has oscillatory solutions in which the amplitude of a is proportional to $y^{-\frac{1}{4}}$. This mild instability probably has little or no practical consequence.

On the the hand, it is clear from the theory that the bubble may become highly unstable to nonspherical disturbances during stage two because $\ddot{R}$ reaches very large positive values during this rebound phase. The instability appears to manifest itself in several different ways depending on the violence of the collapse and the presence of other boundaries. All vapor bubbles that collapse to a size orders of magnitude smaller than their maximum size inevitably emerge from that collapse as a cloud of smaller bubbles rather than a single vapor bubble. This fragmentation could be caused by a single microjet as described below, or it could be due to a spherical harmonic disturbance of higher order. The behavior of collapsing bubbles that are predominantly gas filled (or bubbles whose collapse is thermally inhibited) is less certain since the lower values of $\ddot{R}$ in those cases make the instability weaker and, in some cases, could imply spherical stability. Thus acoustically excited cavitation bubbles that contain substantial gas often remain spherical during their rebound phase. In other instances the instability is sufficient to cause fragmentation. Several examples of fragmented and highly distorted bubbles emerging from the rebound phase are shown in Figure 3.3. These are from the experiments of Frost and Sturtevant (1986), in which the thermal effects are substantial.

A dominant feature in the collapse of many vapor bubbles is the development of a reentrant jet (the $n = 2$ mode) due to an asymmetry such as the presence of a nearby solid boundary. Such an asymmetry causes one side of the bubble to accelerate inward more rapidly than the opposite side and this results in the development of a high-speed re-entrant microjet which penetrates the bubble. Such microjets were first observed experimentally by Naude and Ellis (1961) and Benjamin and Ellis (1966). Of particular interest for cavitation damage is the fact that a nearby solid boundary will cause a microjet directed toward that boundary. Figure 3.4, from Benjamin and Ellis (1966), shows the initial formation of the microjet directed at a nearby wall. Other asymmetries, even gravity, can cause the formation of these reentrant microjets. Figure 3.5 is one of the very first, if not the first, photographs taken showing the result of a gravity-produced upward jet

(a) (b) (c)

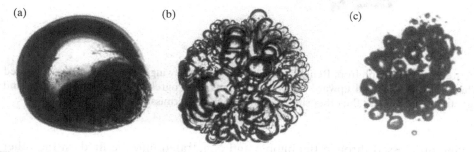

Figure 3.3 Photographs of an ether bubble in glycerine before (left) and after (center) a collapse and rebound. The cloud on the right is the result of a succession of collapse and rebound cycles. Reproduced from Frost and Sturtevant (1986) with the permission of the authors.

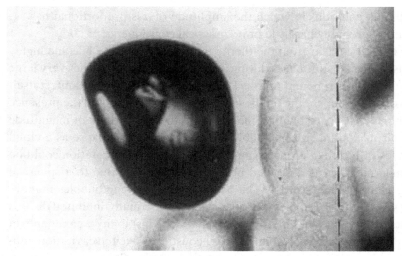

Figure 3.4 Photograph of a collapsing bubble showing the initial development of the reentrant microjet caused by a solid but transparent wall whose location is marked by the dotted line. From Benjamin and Ellis (1966) reproduced with permission of the first author.

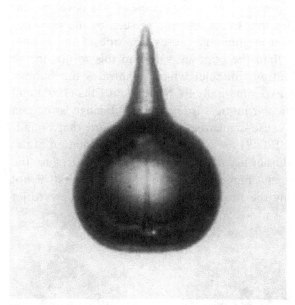

Figure 3.5 Photograph from Benjamin and Ellis (1966) showing the protuberence generated when a gravity-induced upward-directed reentrant jet progresses through the bubble and penetrates the fluid on the other side. Reproduced with permission of the first author.

having progressed through the bubble and penetrated into the fluid on the other side thus creating the spiky protuberance. Indeed, the upward inclination of the wall-induced reentrant jet in Figure 3.4 is caused by gravity. Figure 3.6 presents a

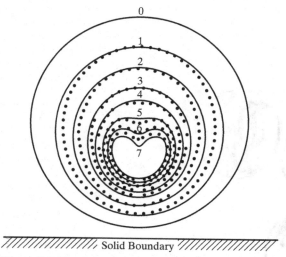

0
1
2
3
4
5
6
7

/////////////// Solid Boundary ///////////////

Figure 3.6 The collapse of a cavitation bubble close to a solid boundary in a quiescent liquid. The theoretical shapes of Plesset and Chapman (1971) (solid lines) are compared with the experimental observations of Lauterborn and Bolle (1975) (points). Figure adapted from Plesset and Prosperetti (1977).

comparison between the reentrant jet development in a bubble collapsing near a solid wall as observed by Lauterborn and Bolle (1975) and as computed by Plesset and Chapman (1971).

Another asymmetry that can cause the formation of a reentrant jet is the proximity of other, neighboring bubbles in a finite cloud of bubbles. Then, as Chahine and Duraiswami (1992) have shown in their numerical calculations, the bubbles on the outer edge of such a cloud will tend to develop jets directed toward the center of the cloud; an example is shown in Figure 3.7. Other manifestations include a bubble collapsing near a free surface, that produces a reentrant jet directed *away* from the free surface (Chahine 1977). Indeed, there exists a critical surface flexibility separating the circumstances in which the reentrant jet is directed away from rather than toward the surface. Gibson and Blake (1982) demonstrated this experimentally and analytically and suggested flexible coatings or liners as a means of avoiding cavitation damage. It might also be noted that depth charges rely for their destructive power on a reentrant jet directed toward the submarine upon the collapse of the explosively generated bubble.

Many other experimentalists have subsequently observed reentrant jets (or "microjets") in the collapse of cavitation bubbles near solid walls. The progress of events seems to differ somewhat depending on the initial distance of the bubble center from the wall. When the bubble is initially spherical but close to the wall, the typical development of the microjet is as illustrated in Figure 3.8, a series of photographs taken by Tomita and Shima (1990). When the bubble is further away from the wall, the later events are somewhat different; another set of photographs taken by Tomita and Shima (1990) is included as Figure 3.9 and shows the formation of two

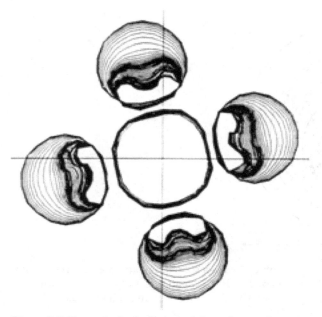

Figure 3.7 Numerical calculation of the collapse of a group of five bubbles showing the development of inward-directed reentrant jets on the outer four bubbles. From Chahine and Duraiswami (1992) reproduced with permission of the authors.

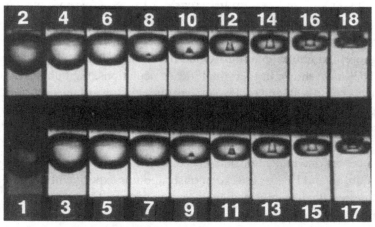

Figure 3.8 Series of photographs showing the development of the microjet in a bubble collapsing very close to a solid wall (at top of frame). The interval between the numbered frames is 2 μs and the frame width is 1.4 *mm*. From Tomita and Shima (1990), reproduced with permission of the authors.

toroidal vortex bubbles (frame 11) after the microjet has completed its penetration of the original bubble. Furthermore, the photographs of Lauterborn and Bolle (1975) in which the bubbles are about a diameter from the wall, show that the initial collapse is quite spherical and that the reentrant jet penetrates the fluid between the bubble and the wall as the bubble is rebounding from the first collapse. At this stage

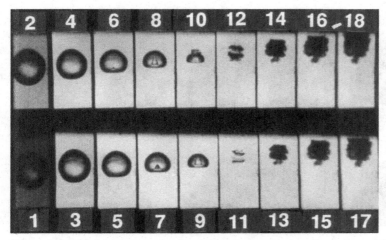

Figure 3.9 A series of photographs similar to the previous figure but with a larger separation from the wall. From Tomita and Shima (1990), reproduced with permission of the authors.

the appearance is very similar to Figure 3.5 but with the protuberance directed at the wall.

On the other hand, when the initial bubble is much closer to the wall and collapse begins from a spherical cap shape, the photographs (for example, Shima et al. (1981) or Kimoto (1987)) show a bubble that "pancakes" down toward the surface in a manner illustrated by Figure 3.10 taken from Benjamin and Ellis (1966). In these circumstances it is difficult to observe the microjet.

The reentrant jet phenomenon in a quiescent fluid has been extensively studied analytically as well as experimentally. Plesset and Chapman (1971) numerically calculated the distortion of an initially spherical bubble as it collapsed close to a solid boundary and, as Figure 3.6 demonstrates, their profiles are in good agreement with the experimental observations of Lauterborn and Bolle (1975). Blake and Gibson (1987) review the current state of knowledge, particularly the analytical methods for solving for bubbles collapsing near a solid or a flexible surface.

When a bubble in a quiescent fluid collapses near a wall, the reentrant jets reach high speeds quite early in the collapse process and long before the volume reaches a size at which, for example, liquid compressibility becomes important (see Section 3.2). The speed of the reentrant jet, U_J, at the time it impacts the opposite surface of the bubble has been shown to be given by

$$U_J = \xi(\Delta p/\rho_L)^{1/2} \tag{3.9}$$

where ξ is a constant and Δp is the difference between the remote pressure, which would maintain the bubble at equilibrium at its maximum or initial radius, and the remote pressure present during collapse. Gibson (1968) found that $\xi = 7.6$ fit his experimental observations; Blake and Gibson (1987) indicate that ξ is a function of ratio, ϱ, of the initial distance of the bubble center from the wall to the initial radius

Figure 3.10 Series of photographs of a hemispherical bubble collapsing against a wall showing the "pancaking" mode of collapse. Four groups of three closely spaced photographs beginning at top left and ending at the bottom right. From Benjamin and Ellis (1966) reproduced with permission of the first author.

and that $\xi = 11.0$ for $\varrho = 1.5$ and $\xi = 8.6$ for $\varrho = 1.0$. Voinov and Voinov (1975) found that the value of ξ could be as high as 64 if the initial bubble had a slightly eccentric shape.

Whether the bubble is fissioned due to the disruption caused by the microjet or by the effects of the stage two instability, many of the experimental observations of bubble collapse (for example, those of Kimoto 1987) show that a bubble emerges from the first rebound not as a single bubble but as a cloud of smaller bubbles. Unfortunately, the events of the last moments of collapse occur so rapidly that the experiments do not have the temporal resolution neccessary to show the details of this fission process. The subsequent dynamical behavior of the bubble cloud may be different from that of a single bubble. For example, the damping of the rebound and collapse cycles is greater than for a single bubble.

Finally, it is important to emphasize that virtually all of the observations described above pertain to bubble collapse in an otherwise quiescent fluid. A bubble that grows and collapses in a flow is subject to other deformations that can significantly alter the noise and damage potential of the collapse process. In Chapter 7 this issue will be addressed further.

3.6 Cavitation Damage

Perhaps the most ubiqitous engineering problem caused by cavitation is the material damage that cavitation bubbles can cause when they collapse in the vicinity of a solid surface. Consequently, this subject has been studied quite intensively for many years (see, for example, ASTM 1967; Thiruvengadam 1967, 1974; Knapp, Daily, and Hammitt 1970). The problem is a difficult one because it involves complicated unsteady flow phenomena combined with the reaction of the particular material of which the solid surface is made. Though there exist many empirical rules designed to help the engineer evaluate the potential cavitation damage rate in a given application, there remain a number of basic questions regarding the fundamental mechanisms involved.

In the preceding sections, we have seen that cavitation bubble collapse is a violent process that generates highly localized, large-amplitude shock waves (Section 3.2) and microjets (Section 3.5) in the fluid at the point of collapse. When this collapse occurs close to a solid surface, these intense disturbances generate highly localized and transient surface stresses. Repetition of this loading due to repeated collapses causes local surface fatigue failure and the subsequent detachment or flaking off of pieces of material. This is the generally accepted explanation of cavitation damage. It is also consistent with the metallurgical evidence of damage in harder materials. Figure 3.11 is a typical photograph of localized cavitation damage on a pump blade. It usually has the jagged, crystalline appearance consistent with fatigue failure and is usually fairly easy to distinguish from the erosion due to solid particles, which has a much smoother appearance. With iron or steel, the effects of corrosion often enhance the speed of cavitation damage.

Parenthetically it should be noted that pits caused by individual bubble collapses are often observed with soft materials, and the relative ease with which this process can be studied experimentally has led to a substantial body of research evidence for soft materials. Much of this literature implies that the microjets cause the individual pits. However, it does not neccessarily follow that the same mechanism causes the damage in harder materials.

Indeed, the issue of whether cavitation damage is caused by microjets or by shock waves or by both has been debated for many years. In the 1940s and 1950s the focus was on the shock waves generated by spherical bubble collapse. When the phenomenon of the microjet was first observed by Naude and Ellis (1961) and Benjamin and Ellis (1966), the focus shifted to studies of the impulsive pressures generated by these jets. But, even after the disruption caused by the microjet, one is left with a remnant cloud of small bubbles that will continue to collapse collectively.

Figure 3.11 Photograph of typical cavitation damage on the blade of a mixed flow pump.

Though no longer a single bubble, this remnant cloud will still exhibit the same qualitative dynamic behavior, including the possible production of a shock wave following the point of minimum cavity volume. Two important research efforts in Japan then shifted the focus back to the remnant cloud shock. First Shima et al. (1983) used high speed Schlieren photography to show that a spherical shock wave was indeed generated by the remnant cloud at the instance of minimum volume. Figure 3.12 shows a series of photographs of a collapsing bubble along with the corresponding pressure trace. The instant of minimum volume is between frames 6 and 7, and the trace clearly shows the peak pressure occuring at that instant. When combined with the Schlieren photographs showing a spherical shock being gener- ated at this instant, this seemed to relegate the microjet to a subsidiary role. About the same time, Fujikawa and Akamatsu (1980) used a photoelastic material so that they could simultaneously observe the stresses in the solid and measure the acoustic pulses. Using the first collapse of a bubble as the trigger, Fujikawa and Akamatsu employed a variable time delay to take photographs of the stress state in the solid at various instants relative to the second collapse. They simultaneously recorded the pressure in the liquid and were able to confirm that the impulsive stresses in the material were initiated at the same moment as the acoustic pulse (to within about 1 μs). They also conclude that this corresponded to the instant of minimum volume and that the waves were not produced by the microjet.

However, in a later investigation, Kimoto (1987) was able to observe stress pulses that resulted both from microjet impingement and from the remnant cloud

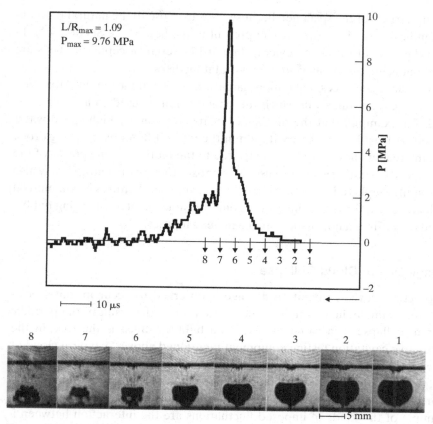

Figure 3.12 Series of photographs of a cavitation bubble collapsing near a wall along with the characteristic wall pressure trace. The time corresponding to each photograph is marked by a number on the trace. From Shima, Takayama, Tomita, and Ohsawa (1983) reproduced with permission of the authors.

collapse shock. Typically, the impulsive pressures from the latter are 2 to 3 times larger than those due to the microjet, but it would seem that both may contribute to the impulsive loading of the surface.

For detailed experimental evaluation of the comparative susceptibility of various materials to cavitation damage, the reader is referred to Knapp, Daily, and Hammitt (1970). Standard devices have been been used to evaluate these comparative susceptibilities. The most common consists of a device that oscillates a specimen in a liquid, producing periodic growth and collapse of cavitation bubbles on the face of the specimen. The tests are conducted over many hours with regular weighing to determine the weight loss. It transpires that the rate of loss of material is not constant. Causes suggested for changes in the rate of loss of material include time constants associated with the fatigue process and the fact that an irregular, damaged surface may produce an altered pattern of cavitation. Commonly, these test cells are operated by a magnetostrictive device in order to achieve the standard frequencies of 5 *kHz* or, sometimes, 20 *kHz*. These frequencies cause the largest cavitating bubble clouds on the surface of the specimen because they are close to the natural frequencies of a significant fraction of the nuclei present in the liquid (see Section 4.2). In addition to the magnetostrictive devices, standard material susceptibility tests are also carried out using cavitating venturis and rotating disks.

In most practical devices, cavitation damage is a very undesirable. However, there are some circumstances in which the phenomenon is used to advantage. It is believed, for example, that the mechanics of rock-cutting by high speed water jets is caused, at least in part, by cavitation in the jet as it flows over a rough rock surface. Many readers have also been subjected to the teeth-cleaning power of the small, high-speed cavitating water jets used by dentists; those with dentures may also have successfully employed acoustic cavitation to clean their dentures in commercial acoustic cleaners. On the other side of the coin, the violence of a collapsing bubble is suspected of causing major tissue damage in head injuries.

3.7 Damage Due to Cloud Collapse

In many practical devices cavitation damage is observed to occur in quite localized areas, for example, in a pump impeller. Often this is the result of the periodic and coherent collapse of a cloud of cavitation bubbles. Such is the case in the magnetostrictive cavitation testing equipment mentioned above. A typical cloud of bubbles generated by such acoustic means is shown in Figure 3.13. In other hydraulic machines, the periodicity may occur naturally as a result of regular shedding of cavitating vortices, or it may be a response to a periodic disturbance imposed on the flow. Example of the kinds of imposed fluctuations are the interaction between a row of rotor vanes and a row of stator vanes in a pump or turbine or the interaction

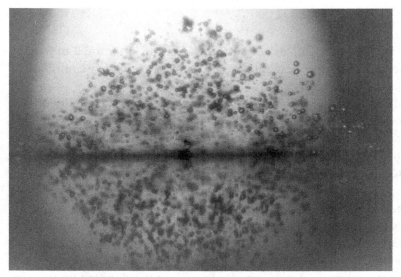

Figure 3.13 Photograph of a transient cloud of cavitation bubbles generated acoustically. From Plesset and Ellis (1955).

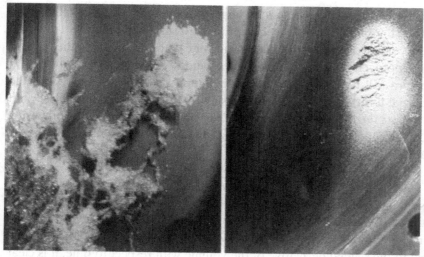

Figure 3.14 Axial views from the inlet of the cavitation and cavitation damage on the hub or base plate of a centrifugal pump impeller. The two photographs are of the same area, the one on the left showing the typical cavitation pattern during flow and the one on the right the typical cavitation damage. Parts of the blades can be seen in the upper left and lower right corners. Relative to these blades, the flow proceeds from the lower left to the upper right. The leading edge of the blade in upper left is just outside the field of view on the left. Reproduced from Soyama, Kato, and Oba (1992) with permission of the authors.

between a ship's propeller and the nonuniform wake behind the ship. In almost all such cases the coherent collapse of the cloud can cause much more intense noise and more potential for damage than in a similar nonfluctuating flow. Consequently the damage is most severe on the solid surface close to the location of cloud collapse. An example of this phenomenon is included in Figure 3.14 taken from Soyama, Kato, and Oba (1992). In this instance clouds of cavitation are being shed from the leading edge of a centrifugal pump blade and are collapsing in a specific location, as suggested by the pattern of cavitation in the left-hand photograph. This leads to the localized damage shown in the right-hand photograph.

At the time of writing, a number of research efforts are focusing on the dynamics of cavitation clouds. Later, in Section 6.10, we analyze some of the basic dynamics of spherical bubble clouds and show that the interaction between bubbles lead to a coherent dynamics of the cloud, including natural frequencies that can be much smaller than the natural frequencies of individual bubbles. These studies suggest that the coherent collapse can be more violent than that of individual bubbles. However, a complete explanation for the increase in the noise and damage potential does not yet exist.

3.8 Cavitation Noise

The violent and catastrophic collapse of cavitation bubbles results in the production of noise as well as the possibility of material damage to nearby solid surfaces. The noise is a consequence of the momentary large pressures that are generated when the contents of the bubble are highly compressed. Consider the flow in the liquid caused by the volume displacement of a growing or collapsing cavity. In the far field the flow will approach that of a simple source, and it is clear that Equation (2.7) for the pressure will be dominated by the first term on the right-hand side (the unsteady inertial term) since it decays more slowly with radius, r, than the second term. If we denote the time-varying volume of the cavity by $V(t)$ and substitute using Equation (2.2), it follows that the time-varying component of the pressure in the far field is given by

$$p_a = \frac{\rho_L}{4\pi \mathcal{R}} \frac{d^2 V}{dt^2} \tag{3.10}$$

where p_a is the radiated acoustic pressure and we denote the distance, r, from the cavity center to the point of measurement by $\mathcal{R}$ (for a more thorough treatment see Dowling and Ffowcs Williams 1983 and Blake 1986b). Since the noise is directly proportional to the second derivative of the volume with respect to time, it is clear that the noise pulse generated at bubble collapse occurs because of the very large and positive values of $d^2 V/dt^2$ when the bubble is close to its minimum size. It is conventional (see, for example, Blake 1986b) to present the sound level using a root mean square pressure or *acoustic* pressure, p_s, defined by

$$p_s^2 = \overline{p_a^2} = \int_0^\infty \mathcal{G}(f) df \tag{3.11}$$

and to represent the distribution over the frequency range, f, by the spectral density function, $\mathcal{G}(f)$.

The crackling noise that accompanies cavitation is one of the most evident characteristics of this phenomenon to the researcher or engineer. The onset of cavitation is often detected first by this noise rather than by visual observation of the bubbles. Moreover, for the practical engineer it is often the primary means of detecting cavitation in devices such as pumps and valves. Indeed, several empirical methods have been suggested that estimate the rate of material damage by measuring the noise generated (for example, Lush and Angell 1984).

The noise due to cavitation in the orifice of a hydraulic control valve is typical, and spectra from such an experiment are presented in Figure 3.15. The lowest curve at $\sigma = 0.523$ represents the turbulent noise from the noncavitating flow. Below the incipient cavitation number (about 0.523 in this case) there is a dramatic increase in the noise level at frequencies of about 5 *kHz* and above. The spectral peak between 5 *kHz* and 10 *kHz* corresponds closely to the expected natural frequencies of the nuclei present in the flow (see Section 4.2).

Most of the analytical approaches to cavitation noise build on knowledge of the dynamics of collapse of a single bubble. Fourier analyses of the radiated acoustic pressure due to a single bubble were first visualized by Rayleigh (1917) and implemented by Mellen (1954) and Fitzpatrick and Strasberg (1956). In considering such Fourier analyses, it is convenient to nondimensionalize the frequency by the typical time span of the whole event or, equivalently, by the collapse time, t_{TC},

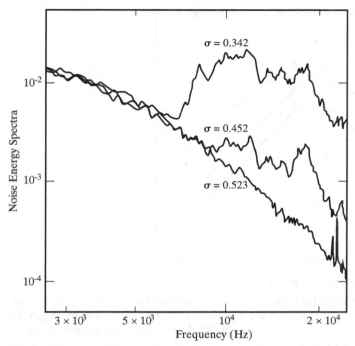

Figure 3.15 Acoustic power spectra from a model spool valve operating under noncavitating ($\sigma = 0.523$) and cavitating ($\sigma = 0.452$ and 0.342) conditions (from the investigation of Martin et al. 1981).

given by Equation (2.40). Now consider the frequency content of $\mathcal{G}(f)$ using the dimensionless frequency, ft_{TC}. Since the volume of the bubble increases from zero to a finite value and then returns to zero, it follows that for $ft_{TC} < 1$ the Fourier transform of the volume is independent of frequency. Consequently d^2V/dt^2 will be proportional to f^2 and therefore $\mathcal{G}(f) \propto f^4$ (see Fitzpatrick and Strasberg 1956). This is the origin of the left-hand asymptote in Figure 3.16. The behavior at intermediate frequencies for which $ft_{TC} > 1$ has been the subject of more speculation and debate. Mellen (1954) and others considered the typical equations governing the collapse of a spherical bubble in the absence of thermal effects and noncondensable gas (Equation (2.36)) and concluded that, since the velocity $\dot{R} \propto R^{-\frac{3}{2}}$, it follows that $R \propto t^{\frac{2}{5}}$. Therefore the Fourier transform of d^2V/dt^2 leads to the asymptotic behavior $\mathcal{G}(f) \propto f^{-\frac{2}{5}}$. The error in this analysis is the neglect of the noncondensable gas. When this is included and when the collapse is sufficiently advanced, the last term in the square brackets of Equation (2.36) becomes comparable with the previous terms. Then the behavior is quite different from $R \propto t^{\frac{2}{5}}$. Moreover, the values of d^2V/dt^2 are much larger during this rebound phase, and therefore the frequency content of the rebound phase will dominate the spectrum. It is therefore not surprising that the $f^{-\frac{2}{5}}$ is not observed in practice. Rather, most of the experimental results seem to exhibit an intermediate frequency behavior like f^{-1} or f^{-2}. Jorgensen (1961) measured the noise from submerged, cavitating jets and found a behavior like f^{-2} at

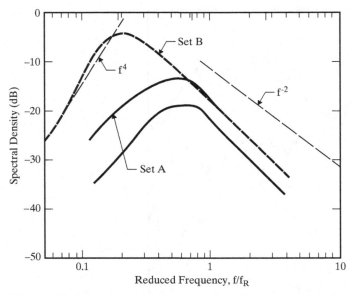

Figure 3.16 Acoustic power spectra of the noise from a cavitating jet. Shown are mean lines through two sets of data constructed by Blake and Sevik (1982) from the data by Jorgensen (1961). Typical asymptotic behaviors are also indicated. The reference frequency, f_R, is $[p_\infty/\rho_L D^2]^{\frac{1}{2}}$ where D is the jet diameter.

the higher frequencies (see Figure 3.16). However, most of the experimental data for cavitating bodies or hydrofoils exhibit a weaker decay. The data by Arakeri and Shangumanathan (1985) from cavitating headform experiments show a very consistent f^{-1} trend over almost the entire frequency range, and very similar results have been obtained by Ceccio and Brennen (1991) (see Figure 3.20). Though the data of Blake et al. (1977) for a cavitating hydrofoil exhibit some consistent peaks, the overall trend in their data is also consistent with f^{-1}. This is also the asymptotic behavior exhibited at higher frequencies by the data of Barker (1973) for a cavitating foil.

Several authors have also analyzed the effects of the compressibility of the liquid. Mellen (1954) and Fitzpatrick and Strasberg (1956) conclude that this causes faster decay like f^{-2} above some critical frequency, though this has not been clearly demonstrated experimentally. In Figure 3.16 typical functional behaviors of $\mathcal{G}(f)$ have been included in a graph showing measurements of some of the noise from a cavitating jet taken by Jorgensen (1961).

The peaks in many of the spectra of the noise from cavitating flows (for example those of Figures 3.15 and 3.16) tend to lower frequencies as the cavitation number decreases, primarily because of an increase in the amplitude of the higher frequencies. This trend is further illustrated by the data for cavitating jets presented in Figure 3.17. Given some form for the asymptotes at high and low frequency and some functional behavior for the location of the peak frequency, it is not unreasonable to seek to reduce the measured spectra for a given type of cavitating flow to some universal form. Arakeri and Shangumanathan (1985) were able to reduce the

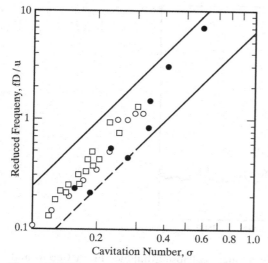

Figure 3.17 Frequency of the peak in the acoustic spectra as a function of cavitation number. Data for cavitating jets from Franklin and McMillan (1984). The jet diameter and mean velocity are denoted by D and U.

noise spectra from cavitating headform experiments to a single band provided the bubble population was low enough to eliminate bubble/bubble interaction. Blake (1986a) has attempted a similar task for the various kinds of cavitation that can occur on a propeller, since the ability to scale from model tests to the prototype is important in that context. There remain, however, a number of unresolved issues that seem to demand a closer examination of the basic mechanics of noise production even in the absence of bubble/bubble interactions.

Recently Ceccio and Brennen (1991) have recorded the noise from individual cavitation bubbles in a flow; a typical acoustic signal from their experiments is reproduced in Figure 3.18. The large positive pulse at about $450\,\mu s$ corresponds to the first collapse of the bubble. This first pulse in Figure 3.18 is followed by some facility-dependent oscillations and by a second pulse at about $1100\,\mu s$. This corresponds to the second collapse, which follows the rebound from the first collapse. Further rebounds are possible but were not observed in these experiments.

A good measure of the magnitude of the collapse pulse is the acoustic impulse, I, defined as the area under the pulse or

$$I = \int_{t_1}^{t_2} p_a dt \qquad (3.12)$$

where t_1 and t_2 are times before and after the pulse at which p_a is zero. For later purposes we also define a dimensionless impulse, I^*, as

$$I^* = 4\pi I\mathcal{R}/\rho_L U_\infty R_H^2 \qquad (3.13)$$

where U_∞ and R_H are the reference velocity and length in the flow. The average acoustic impulses for individual bubble collapses on two axisymmetric headforms

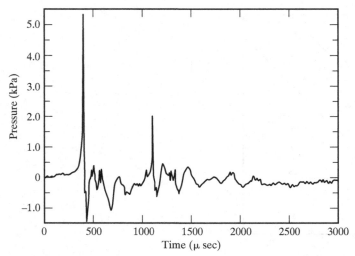

Figure 3.18 A typical acoustic signal from a single collapsing bubble. From Ceccio and Brennen (1991).

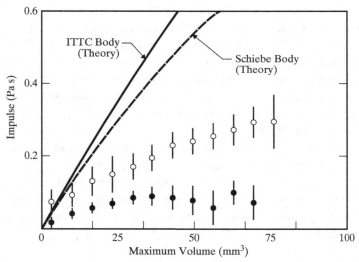

Figure 3.19 Comparison of the acoustic impulse, I, produced by the collapse of a single cavitation bubble on two axisymmetric headforms as a function of the maximum volume prior to collapse. Open symbols: average data for Schiebe headform; closed symbols: ITTC headform; vertical lines indicate one standard deviation. Also shown are the corresponding results from the solution of the Rayleigh-Plesset equation. From Ceccio and Brennen (1991).

(ITTC and Schiebe headforms) are compared in Figure 3.19 with impulses predicted from integration of the Rayleigh-Plesset equation. Since these theoretical calculations assume that the bubble remains spherical, the discrepancy between the theory and the experiments is not too surprising. Indeed one interpretation of Figure 3.19 is that the theory can provide an order of magnitude estimate and an upper bound

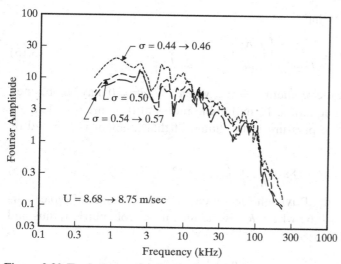

Figure 3.20 Typical spectra of noise from bubble cavitation for various cavitation numbers. From Ceccio and Brennen (1991).

on the noise produced by a single bubble. In actuality, the departure from sphericity produces a less focused collapse and therefore less noise.

Typical spectra showing the frequency content in single bubble noise are included in Figure 3.20. If the events are randomly distributed in time (see below), this would also correspond to the overall cavitation noise spectrum. These spectra exhibit the previously mentioned f^{-1} behavior for the range $1 \rightarrow 50$ *kHz*; the rapid decline at about 80 *kHz* represents the limit of the hydrophone used to make these measurements.

The next step is to consider the synthesis of cavitation noise from the noise produced by individual cavitation bubbles or events. This is a fairly simple matter provided the events can be considered to occur randomly in time. At low nuclei population densities the evidence suggests that this is indeed the case (see, for example, Morozov 1969). Baiter, Gruneis, and Tilmann (1982) have explored the consequences of the departures from randomness that could occur at larger bubble population densities. Here, we limit the analysis to the case of random events. Then, if the impulse produced by each event is denoted by I and the number of events per unit time is denoted by $\dot{N}_E$, the sound pressure level, p_S, will be given by

$$p_S = I \dot{N}_E \tag{3.14}$$

Consider the scaling of cavitation noise that is implicit in this construct. We shall omit some factors of proportionality for the sake of clarity, so the results are only intended as a qualitative guide.

Both the experimental results and the analysis based on the Rayleigh-Plesset equation indicate that the nondimensional impulse produced by a single cavitation event is strongly correlated with the maximum volume of the bubble prior to collapse and is almost independent of the other flow parameters. It follows from

Equations (3.10) and (3.12) that

$$I^* = \frac{1}{U_\infty R_H^2}\left[\left(\frac{dV}{dt}\right)_{t_2} - \left(\frac{dV}{dt}\right)_{t_1}\right] \tag{3.15}$$

and the values of dV/dt at the moments $t = t_1, t_2$ when $d^2V/dt^2 = 0$ may be obtained from the Rayleigh-Plesset equation. If the bubble radius at the time t_1 is denoted by R_X and the coefficient of pressure in the liquid at that moment is denoted by C_{px}, then

$$I^* \approx 8\pi \left(\frac{R_X}{R_H}\right)^2 (C_{px} - \sigma)^{\frac{1}{2}} \tag{3.16}$$

Numerical integrations of the Rayleigh-Plesset equation for a range of typical circumstances yield $R_X/R_M \approx 0.62$ where R_M is the maximum volumetric radius and that $(C_{px} - \sigma) \propto R_M/R_H$ so that

$$I^* \approx \beta \left(\frac{R_M}{R_H}\right)^{\frac{5}{2}} \tag{3.17}$$

The aforementioned integrations of the Rayleigh-Plesset equation yield a factor of proportionality, β, of about 35. Moreover, the upper envelope of the experimental data of which Figure 3.19 is a sample appears to correspond to a value of $\beta \approx 4$. We note that a quite similar relation between I^* and R_M/R_H emerges from the analysis by Esipov and Naugol'nykh (1973) of the compressive sound wave generated by the collapse of a gas bubble in a compressible liquid. Indeed, the compressibility of the liquid does not appear to affect the acoustic impulse significantly.

From the above relations, it follows that

$$I \approx \frac{\beta}{12}\rho_L U_\infty R_M^{\frac{5}{2}}/\mathcal{R}R_H^{\frac{1}{2}} \tag{3.18}$$

Consequently, the evaluation of the impulse from a single event is completed by some estimate of R_M. Previously (Section 2.5) we evaluated R_M and showed it to be independent of U_∞ for a given cavitation number. In that case I is linear with U_∞.

The event rate, $\dot{N}_E$, can be considerably more complicated to evaluate than might at first be thought. If all the nuclei flowing through a certain, known streamtube (say with a cross-sectional area in the upstream flow of A_N) were to cavitate similarly, then the result would be

$$\dot{N}_E = NA_N U_\infty \tag{3.19}$$

where N is the nuclei concentration (number/unit volume) in the incoming flow. Then it follows that the acoustic pressure level resulting from substituting Equations (3.19), (3.18) and (2.52) into Equation (3.14) becomes

$$p_S \approx \frac{\beta}{3}\rho_L U_\infty^2 A_N N R_H^2 (-\sigma - C_{pmin})^{\frac{5}{2}}/\mathcal{R} \tag{3.20}$$

where we have omitted some of the constants of order unity. For the relatively simple flows considered here, Equation (3.20) yields a sound pressure level that scales with U_∞^2 and with R_H^4 because $A_N \propto R_H^2$. This scaling with velocity does correspond roughly to that which has been observed in some experiments on traveling bubble cavitation, for example, those of Blake, Wolpert, and Geib (1977) and Arakeri and Shangumanathan (1985). The former observe that $p_S \propto U_\infty^m$ where $m = 1.5$ to 2. There are, however, a number of complicating factors that can alter these scaling relationships. First, as we have discussed earlier in Section 2.5, only those nuclei larger than a certain critical size, R_C, will actually grow to become cavitation bubbles. Since R_C is a function of both σ and the velocity, U_∞, this means that the effective N will be a function of R_C and U_∞. Since R_C decreases as U_∞ increases, this would tend to produce powers, m, somewhat greater than 2. But it is also the case that, in any experimental facility, N will typically change with U_∞ in some facility-dependent manner. Often this will cause N to decrease with U_∞ at constant σ (since N will typically decrease with increasing tunnel pressure), and this effect would then produce values of m that are less than 2.

Different scaling laws will apply when the cavitation is generated by turbulent fluctuations such as in a turbulent jet (see, for example, Ooi 1985 and Franklin and McMillan 1984). Then the typical tension experienced by a nucleus as it moves along a disturbed path in a turbulent flow is very much more difficult to estimate. Consequently, the models for the sound pressure due to cavitation and the scaling of that sound with velocity are less well understood.

When the population of bubbles becomes sufficiently large, the radiated noise will begin to be affected by the interactions between the bubbles. In Chapters 6 and 7 we discuss some analyses and some of the consequences of these interactions in clouds of cavitating bubbles. There are also a number of experimental studies of the noise from the collapse of cavitating clouds, for example, that of Bark and van Berlekom (1978).

3.9 Cavitation Luminescence

Though highly localized both temporally and spatially, the extremely high temperatures and pressures that can occur in the noncondensable gas during collapse are believed to be responsible for the phenomenon known as luminescence, the emission of light that is observed during cavitation bubble collapse. The phenomenon was first observed by Marinesco and Trillat (1933), and a number of different explanations were advanced to explain the emissions. The fact that the light was being emitted at collapse was first demonstrated by Meyer and Kuttruff (1959). They observed cavitation on the face of a rod oscillating magnetostrictively and correlated the light with the collapse point in the growth-and-collapse cycle. The balance of evidence now seems to confirm the suggestion by Noltingk and Neppiras (1950) that the phenomenon is caused by the compression and adiabatic heating of the noncondensable gas in the collapsing bubble. As we discussed previously in Sections 2.4 and 3.4, temperatures of the order of $6000°K$ can be anticipated on the basis of uniform compression of the noncondensable gas; the same calculations suggest that these high temperatures will last for only a fraction of a microsecond. Such conditions would indeed explain the emission of light. Indeed, the measurements of the spectrum of sonoluminescence by Taylor and Jarman (1970), Flint and Suslick (1991), and others suggest a temperature of about $5000K$. However, some recent experiments by Barber and Putterman (1991) indicate much higher temperatures and even shorter emission durations of the order of picoseconds. Speculations on the explanation for these observations have centered on the suggestion by Jarman (1960) that the collapsing bubble forms a spherical, inward-propagating shock in the gas contents of the bubble and that the focusing of the shock at the center of the bubble is an important reason for the extremely high apparent "temperatures" associated with the sonoluminescence radiation. It is, however, important to observe that spherical symmetry is essential for this mechanism to have any significant consequences. One would therefore expect that the distortions caused by a flow would not allow significant shock focusing and would even reduce the effectiveness of the basic compression mechanism.

When it occurs in the context of acoustic cavitation (see Chapter 4), luminescence is called sonoluminescence despite the evidence that it is the cavitation rather than the sound that causes the light emission. Sonoluminescence and the associated chemistry that is induced by the high temperatures and pressures (known as "sonochemistry") have been more thoroughly investigated than the corresponding processes in hydrodynamic cavitation. However, the subject is beyond the scope of this book and the reader is referred to other works such as the book by Young (1989). As one would expect from the Rayleigh-Plesset equation, the surface tension and vapor pressure of the liquid are important in determining the sonoluminescence flux as the data of Jarman (1959) clearly show (see Figure 3.21). Certain aqueous solutes like sodium disulphide seem to enhance the luminescence, though it is not clear that the same mechanism is responsible for the light emission under these circumstances. Sonoluminescence is also strongly dependent on the thermal conductivity of the gas (Hickling 1963, Young 1976), and this is particularly evident with gases like xenon and krypton, which have low thermal conductivities. Clearly then, the conduction of heat in the gas plays an important role in the phenomenon. Therefore, the breakup of the bubble prior to complete collapse might be expected to eliminate the phenomenon completely.

Light emission in a cavitating flow was first investigated by Jarman and Taylor (1964, 1965) who observed luminescence in a cavitating venturi and identified the source as the region of bubble collapse. They also found that an acoustic pressure pulse was associated with each flash of light. The maximum emission was in a band of wavelengths around 5000 angstroms, which is in accord with the many apocryphal accounts of steady or flashing blue light emanating from flowing water. Peterson and Anderson (1967) also conducted experiments with venturis and explored the effects of different, noncondensable gases dissolved in the water. They observe that the emission of light implies blackbody sources with a temperature above $6000°K$. There are, however, other experimenters who found it very difficult to observe any luminescence in a cavitating flow. One suspects that only bubbles that collapse with

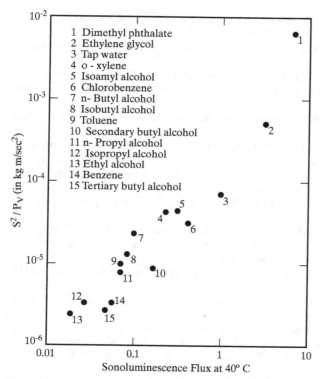

Figure 3.21 The correlation of the sonoluminescence flux with S^2/p_V for data in a variety of liquids. From Jarman (1959).

significant spherical symmetry will actually produce luminescence. Such events may be exceedingly rare in many flows.

The phenomenon of luminescence is not just of academic interest. For one thing there is evidence that it may initiate explosions in liquid explosives (Gordeev et al. 1967). More constructively, there seems to be significant interest in utilizing the chemical-processing potential of the high temperatures and pressures in what is otherwise a benign environment. For example, it is possible to use cavitation to break up harmful molecules in water (Dahi 1982).

References

ASTM. (1967). *Erosion by cavitation or impingement.* Amer. Soc. for Testing and Materials, ASTM STP408.

Akulichev, V.A. (1971). *High intensity ultrasonic fields.* L.D. Rosenberg (ed.), Plenum Press.

Arakeri, V.H. and Shangumanathan, V. (1985). On the evidence for the effect of bubble interference on cavitation noise. *J. Fluid Mech.*, **159**, 131–150.

Baiter, H.-J., Gruneis, F. and Tilmann, P. (1982). An extended base for the statistical description of cavitation noise. *Proc. ASME Int. Symp. on Cavitation Noise*, 93–108.

Barber, B.P. and Putterman, S.J. (1991). Observations of synchronous picosecond sonluminescence. *Nature*, **352**, 318.

Bark, G. and van Berlekom, W.B. (1978). Experimental investigations of cavitation noise. *Proc. 12th ONR Symp. on Naval Hydrodynamics*, 470–493.

Barker, S.J. (1973). Measurements of radiated noise from cavitating hydrofoils. *ASME Cavitation and Polyphase Flow Forum*, 27–30.

Benjamin, T.B. (1958). Pressure waves from collapsing cavities. *Proc. Second ONR Symp. Naval Hyrodynamics*, 207–233.

Benjamin, T.B. and Ellis, A.T. (1966). The collapse of cavitation bubbles and the pressures thereby produced against solid boundaries. *Phil. Trans. Roy. Soc., London, Ser. A*, **260**, 221–240.

Blake, J.R. and Gibson, D.C. (1987). Cavitation bubbles near boundaries. *Ann. Rev. Fluid Mech.*, **19**, 99–124.

Blake, W.K., Wolpert, M.J. and Geib, F.E. (1977). Cavitation noise and inception as influenced by boundary-layer development on a hydrofoil. *J. Fluid Mech.*, **80**, 617–640.

Blake, W.K. and Sevik, M.M. (1982). Recent developments in cavitation noise research. *Proc. ASME Int. Symp. on Cavitation Noise*, 1–10.

Blake, W.K. (1986a). Propeller cavitation noise: the problems of scaling and prediction. *Proc. ASME Int. Symp. on Cavitation Noise*, 89–99.

Blake, W.K. (1986b). *Mechanics of flow-induced sound and vibration.* Academic Press.

Ceccio, S.L. and Brennen, C.E. (1991). Observations of the dynamics and acoustics of travelling bubble cavitation. *J. Fluid Mech.*, **233**, 633–660.

Chahine, G.L. (1977). Interaction between an oscillating bubble and a free surface. *ASME J. Fluids Eng.*, **99**, 709–716.

Chahine, G.L. and Duraiswami, R. (1992). Dynamical interactions in a multibubble cloud. *ASME J. Fluids Eng.*, **114**, 680–686.

Dahi, E. (1982). Perspective of combination of ozone and ultrasound. In *Ozonization Manual for Water and Wastewater Treatment* (editor W.J.Masschelein), John Wiley and Sons.

Dowling, A.P. and Ffowcs Williams, J.E. (1983). *Sound and sources of sound.* Ellis Horwood Ltd. and John Wiley and Sons.

Esipov, I.B. and Naugol'nykh, K.A. (1973). Collapse of a bubble in a compressible liquid. *Akust. Zh.*, **19**, 285–288.

Fitzpatrick, H.M. and Strasberg, M. (1956). Hydrodynamic sources of sound. *Proc. First ONR Symp. on Naval Hydrodynamics*, 241–280.

Flint, E.B. and Suslick, K.S. (1991). The temperature of cavitation. *Science*, **253**, 1397–1399.

Flynn, H.G. (1966). Cavitation dynamics. I, A mathematical formulation. *Acoust. Res. Lab., Harvard Univ.*, Tech. Memo. 50.

Franklin, R.E. and McMillan, J. (1984). Noise generation in cavitating flows, the submerged jet. *ASME J. Fluids Eng.*, **106**, 336–341.

Frost, D. and Sturtevant, B. (1986). Effects of ambient pressure on the instability of a liquid boiling explosively at the superheat limit. *ASME J. Heat Transfer*, **108**, 418–424.

Fujikawa, S. and Akamatsu, T. (1980). Effects of the non-equilibrium condensation of vapour on the pressure wave produced by the collapse of a bubble in a liquid. *J. Fluid Mech.*, **97**, 481–512.

Gibson, D.C. (1968). Cavitation adjacent to plane boundaries. *Proc. Australian Conf. on Hydraulics and Fluid Machinery*, 210–214.

Gibson, D.C. and Blake, J.R. (1982). The growth and collapse of bubbles near deformable surfaces. *Appl. Sci. Res.*, **38**, 215–224.

Gilmore, F.R. (1952). The collapse and growth of a spherical bubble in a viscous compressible liquid. *Calif. Inst. of Tech. Hydrodynamics Lab. Rep. No. 26–4.*

Gordeev, V.E., Serbinov, A.I., and Troshin, Ya.K. (1967). Stimulation of explosions in the collapse of cavitation bubbles in liquid explosives. *Dokl. Akad. Nauk.*, **172**, 383–385.

Herring, C. (1941). Theory of the pulsations of the gas bubble produced by an underwater explosion. *O.S.R.D. Rep. No. 236.*

Hickling, R. (1963). Effects of thermal conduction in sonoluminescence. *J. Acoust. Soc. Am.*, **35**, 967–974.

Hickling, R. and Plesset, M.S. (1964). Collapse and rebound of a spherical bubble in water. *Phys. Fluids*, **7**, 7–14.

Ivany, R.D. and Hammitt, F.G. (1965). Cavitation bubble collapse in viscous, compressible liquids—numerical analysis. *ASME J. Basic Eng.*, **87**, 977–985.

Jarman, P. (1959). Measurements of sonoluminescence from pure liquids and some aqueous solutions. *Proc. Phys. Soc. London*, **73**, 628–640.

Jarman, P. (1960). Sonoluminescence: a discussion. *J. Acoust. Soc. Am.*, **32**, 1459–1462.

Jarman, P. and Taylor, K.J. (1964). Light emisssion from cavitating water. *Brit. J. Appl. Phys.*, **15**, 321–322.

Jarman, P. and Taylor, K.J. (1965). Light flashes and shocks from a cavitating flow. *Brit. J. Appl. Phys.*, **16**, 675–682.

Jorgensen, D.W. (1961). Noise from cavitating submerged jets. *J. Acoust. Soc. Am.*, **33**, 1334–1338.

Keller, J.B. and Kolodner, I.I. (1956). Damping of underwater explosion bubble oscillations. *J. Appl. Phys.*, **27**, 1152–1161.

Kimoto, H. (1987). An experimental evaluation of the effects of a water microjet and a shock wave by a local pressure sensor. *Int. ASME Symp. on Cavitation Res. Facilities and Techniques*, **FED 57**, 217–224.

Kirkwood, J.G. and Bethe, H.A. (1942). The pressure wave produced by an underwater explosion. *O.S.R.D.*, Rep. 588.

Knapp, R.T., Daily, J.W., and Hammitt, F.G. (1970). *Cavitation*. McGraw-Hill, New York.

Lauterborn, W. and Bolle, H. (1975). Experimental investigations of cavitation bubble collapse in the neighborhood of a solid boundary. *J. Fluid Mech.*, **72**, 391–399.

Lush, P.A. and Angell, B. (1984). Correlation of cavitation erosion and sound pressure level. *ASME. J. Fluids Eng.*, **106**, 347–351.

Marinesco, M. and Trillat, J.J. (1933). Action des ultrasons sur les plaques photographiques. *Compt. Rend.*, **196**, 858–860.

Martin, C.S., Medlarz, H., Wiggert, D.C., and Brennen, C. (1981). Cavitation inception in spool valves. *ASME. J. Fluids Eng.*, **103**, 564–576.

Matsumoto, Y. and Watanabe, M. (1989). Nonlinear oscillation of gas bubble with internal phenomena. *JSME Int. J.*, **32**, 157–162.

Mellen, R.H. (1954). Ultrasonic spectrum of cavitation noise in water. *J. Acoust. Soc. Am.*, **26**, 356–360.

Meyer, E. and Kuttruff, H. (1959). Zur Phasenbeziehung zwischen Sonolumineszenz und Kavitations-vorgang bei periodischer Anregung. *Zeit angew. Phys.*, **11**, 325–333.

Morozov, V.P. (1969). Cavitation noise as a train of sound pulses generated at random times. *Sov. Phys. Acoust.*, **14**, 361–365.

Naude, C.F. and Ellis, A.T. (1961). On the mechanism of cavitation damage by non-hemispherical cavities in contact with a solid boundary. *ASME. J. Basic Eng.*, **83**, 648–656.

Noltingk, B.E. and Neppiras, E.A. (1950). Cavitation produced by ultrasonics. *Proc. Phys. Soc., London*, **63B**, 674–685.

Ooi, K.K. (1985). Scale effects on cavitation inception in submerged water jets: a new look. *J. Fluid Mech.*, **151**, 367–390.

Peterson, F.B. and Anderson, T.P. (1967). Light emission from hydrodynamic cavitation. *Phys. Fluids*, **10**, 874–879.

Plesset, M.S. and Ellis, A.T. (1955). On the mechanism of cavitation damage. *Trans. ASME*, 1055–1064.

Plesset, M.S. and Mitchell, T.P. (1956). On the stability of the spherical shape of a vapor cavity in a liquid. *Quart. Appl. Math.*, **13**, No. 4, 419–430.

Plesset, M.S. and Chapman, R.B. (1971). Collapse of an initially spherical vapor cavity in the neighborhood of a solid boundary. *J. Fluid Mech.*, **47**, 283–290.

Plesset, M.S. and Prosperetti, A. (1977). Bubble dynamics and cavitation. *Ann. Rev. Fluid Mech.*, **9**, 145–185.

Prosperetti, A. and Lezzi, A. (1986). Bubble dynamics in a compressible liquid. Part 1. First-order theory. *J. Fluid Mech.*, **168**, 457–478.

Rayleigh, Lord (Strutt, John William). (1917). On the pressure developed in a liquid during the collapse of a spherical cavity. *Phil. Mag.*, **34**, 94–98.

Schneider, A.J.R. (1949). Some compressibility effects in cavitation bubble dynamics. *Ph.D. Thesis, Calif. Inst. of Tech.*

Shima, A., Takayama, K., Tomita, Y., and Miura, N. (1981). An experimental study on effects of a solid wall on the motion of bubbles and shock waves in bubble collapse. *Acustica*, **48**, 293–301.

Shima, A., Takayama, K., Tomita, Y., and Ohsawa, N. (1983). Mechanism of impact pressure generation from spark-generated bubble collapse near a wall. *AIAA J.*, **21**, 55–59.

Soyama,H., Kato, H., and Oba, R. (1992). Cavitation observations of severely erosive vortex cavitation arising in a centrifugal pump. *Proc. Third I. Mech. E. Int. Conf. on Cavitation*, 103–110.

Taylor, K.J. and Jarman, P.D. (1970). The spectra of sonoluminescence. *Aust. J. Phys.*, **23**, 319–334.

Theofanous, T., Biasi, L., Isbin, H.S., and Fauske, H. (1969). A theoretical study on bubble growth in constant and time-dependent pressure fields. *Chem. Eng. Sci.*, **24**, 885–897.

Thiruvengadam, A. (1967). The concept of erosion strength. In *Erosion by cavitation or impingement.* Am. Soc. Testing Mats. STP 408, 22–35.

Thiruvengadam, A. (1974). Handbook of cavitation erosion. *Tech. Rep. 7301-1, Hydronautics, Inc., Laurel, Md.*

Tomita, Y. and Shima, A. (1977). On the behaviour of a spherical bubble and the impulse pressure in a viscous compressible liquid. *Bull. JSME*, **20**, 1453–1460.

Tomita, Y. and Shima, A. (1990). High-speed photographic observations of laser-induced cavitation bubbles in water. *Acustica*, **71**, No. 3, 161–171.

Trilling, L. (1952). The collapse and rebound of a gas bubble. *J. Appl. Phys.*, **23**, 14–17.

Voinov, O.V. and Voinov, V.V. (1975). Numerical method of calculating non-stationary motions of an ideal incompressible fluid with free surfaces. *Sov. Phys. Dokl.*, **20**, 179–180.

Young, F.R. (1976). Sonoluminescence from water containing dissolved gases. *J. Acoust. Soc. Am.*, **60**, 100–104.

Young, F.R. (1989). *Cavitation.* McGraw-Hill Book Company.

4 Dynamics of Oscillating Bubbles

4.1 Introduction

The focus of the two preceding chapters was on the dynamics of the growth and collapse of a single bubble experiencing one period of tension. In this chapter we review the response of a bubble to a continuous, oscillating pressure field. Much of the material comes within the scope of acoustic cavitation, a subject with an extensive literature that is reviewed in more detail elsewhere (Flynn 1964; Neppiras 1980; Plesset and Prosperetti 1977; Prosperetti 1982, 1984; Crum 1979; Young 1989). We include here a brief summary of the basic phenomena.

One useful classification of the subject uses the magnitude of the bubble radius oscillations in response to the imposed fluctuating pressure field. Three regimes can be identified:

1. For very small pressure amplitudes the response is linear. Section 4.2 contains the first step in any linear analysis, the identification of the natural frequency of an oscillating bubble.
2. Due to the nonlinearities in the governing equations, particularly the Rayleigh-Plesset Equation (2.12), the response of a bubble will begin to be affected by these nonlinearities as the amplitude of oscillation is increased. Nevertheless the bubble *may* continue to oscillate stably. Such circumstances are referred to as "stable acoustic cavitation" to distinguish them from those of the third regime described below. Several different nonlinear phenomena can affect stable acoustic cavitation in important ways. Among these are the production of subharmonics, the phenomenon of rectified diffusion, and the generation of Bjerknes forces. Each of these is described in greater detail later in the chapter.
3. Under other circumstances the change in bubble size during a single cycle of oscillation can become so large that the bubble undergoes a cycle of explosive cavitation growth and violent collapse similar to that described in the preceding chapter. Such a response is termed "transient acoustic cavitation" and is distinguished from stable acoustic cavitation by the fact that the bubble radius changes by several orders of magnitude during each cycle.

Though we imply that these three situations follow with increasing amplitude, it is important to note that other factors are important in determining the kind of response that will occur for a given oscillating pressure field. One of the factors is the relationship between the frequency, ω, of the imposed oscillations and the natural frequency, ω_N, of the bubble. Sometimes this is characterized by the relationship between the equilibrium radius of the bubble, R_E, in the absence of pressure oscillations and the size of the hypothetical bubble, R_R, which would resonate at the imposed frequency, ω. Another important factor in determining whether the response is stable or transient is the relationship between the pressure oscillation amplitude, $\tilde{p}$, and the mean pressure, $\bar{p}_\infty$. For example, if $\tilde{p} < \bar{p}_\infty$, the bubble is never placed under tension and will therefore never cavitate. A related factor that will affect the response is whether the bubble is predominantly vapor-filled or gas-filled. Stable oscillations are more likely with predominantly gas-filled bubbles while bubbles which contain mostly vapor will more readily exhibit transient acoustic cavitation.

We begin, however, with a discussion of the small-amplitude, linear response of a bubble to oscillations in the ambient pressure.

4.2 Bubble Natural Frequencies

The response of a bubble to oscillations in the pressure at infinity will now be considered. Initially we shall neglect thermal effects and the influence of liquid compressibility. As discussed in the next section both of these lead to an increase in the damping above that represented by the viscous terms, which are retained. However, both can be approximately represented by increases in the damping or the "effective" viscosity.

Consider the linearized dynamic solution of Equation (2.27) when the pressure at infinity consists of a mean value, $\bar{p}_\infty$, upon which is superimposed a *small* oscillatory pressure of amplitude, $\tilde{p}$, and radian frequency, ω, so that

$$p_\infty = \bar{p}_\infty + Re\{\tilde{p}e^{j\omega t}\} \tag{4.1}$$

The linear dynamic response of the bubble will then be

$$R = R_E[1 + Re\{\varphi e^{j\omega t}\}] \tag{4.2}$$

where R_E is the equilibrium size at the pressure, $\bar{p}_\infty$, and the bubble radius response, φ, will in general be a complex number such that $R_E|\varphi|$ is the amplitude of the bubble radius oscillations. The phase of φ represents the phase difference between p_∞ and R.

For the present we shall assume that the mass of gas in the bubble, m_G, remains constant. Then substituting Equations (4.1) and (4.2) into Equation (2.27), neglecting all terms of order $|\varphi|^2$ and using the equilibrium condition (2.41) one finds

$$\omega^2 - j\omega\frac{4\nu_L}{R_E^2} + \frac{1}{\rho_L R_E^2}\left[\frac{2S}{R_E} - 3kp_{GE}\right] = \frac{\tilde{p}}{\rho_L R_E^2 \varphi} \tag{4.3}$$

where, as before,

$$p_{GE} = \bar{p}_\infty - p_V + \frac{2S}{R_E} = \frac{3m_G T_B K_G}{4\pi R_E^3} \tag{4.4}$$

It follows that for a given amplitude, $\tilde{p}$, the maximum or peak response amplitude occurs at a frequency, ω_P, given by the minimum value of the spectral radius of the left-hand side of Equation (4.3):

$$\omega_P = \left[\frac{(3kp_{GE} - 2S/R_E)}{\rho_L R_E^2} - \frac{8v_L^2}{R_E^4} \right]^{\frac{1}{2}} \tag{4.5}$$

or in terms of $(\bar{p}_\infty - p_V)$ rather than p_{GE}:

$$\omega_P = \left[\frac{3k(\bar{p}_\infty - p_V)}{\rho_L R_E^2} + \frac{2(3k-1)S}{\rho_L R_E^3} - \frac{8v_L^2}{R_E^4} \right]^{\frac{1}{2}} \tag{4.6}$$

At this peak frequency the amplitude of the response is, of course, inversely proportional to the damping:

$$|\varphi|_{\omega=\omega_P} = \frac{\tilde{p}}{4\mu_L \left[\omega_P^2 + \frac{4v_L^2}{R_E^4} \right]^{\frac{1}{2}}} \tag{4.7}$$

It is also convenient for future purposes to define the natural frequency, ω_N, of oscillation of the bubbles as the value of ω_P for zero damping:

$$\omega_N = \left[\frac{1}{\rho_L R_E^2} \left\{ 3k(\bar{p}_\infty - p_V) + 2(3k-1)\frac{S}{R_E} \right\} \right]^{\frac{1}{2}} \tag{4.8}$$

The connection with the stability criterion of Section 2.5 is clear when one observes that no natural frequency exists for tensions $(p_V - \bar{p}_\infty) > 4S/3R_E$ (for isothermal gas behavior, $k = 1$); stable oscillations can only occur about a stable equilibrium.

The peak frequency, ω_P, is an important quantity to consider in any bubble dynamic problem. Note from Equation (4.41) that ω_P is a function only of $(\bar{p}_\infty - p_V), R_E$, and the liquid properties. Typical graphs for ω_P as a function of R_E for several $(\bar{p}_\infty - p_V)$ values are shown in Figures (4.1) and (4.2) for water at $300°K$ ($S = 0.0717$, $\mu_L = 0.000863$, $\rho_L = 996.3$) and for sodium at $800°K$ ($S = 0.15$, $\mu_L = 0.000229$, $\rho_L = 825.8$). As is evident from Equation (4.41), the second and third terms on the right-hand side dominate at very small R_E and the frequency is almost independent of $(\bar{p}_\infty - p_V)$. Indeed, no peak frequency exists below a size equal to about $2v_L^2 \rho_L/S$. For larger bubbles the viscous term becomes negligible and ω_P depends on $(\bar{p}_\infty - p_V)$. If the latter is positive, the natural frequency approaches zero like R_E^{-1}. In the case of tension, $p_V > \bar{p}_\infty$, the peak frequency does not exist above $R_E = R_C$.

It is important to take note of the fact that for the typical nuclei commonly found in water, which lie in range 1 to 100 μm, the natural frequencies are of the order, 5 to 25 kHz. This has several important practical consequences. First, if one

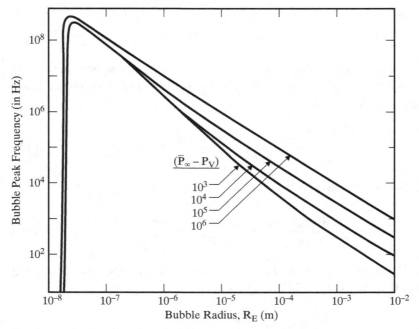

Figure 4.1 Bubble resonant frequency in water at $300°K$ ($S = 0.0717$, $\mu_L = 0.000863$, $\rho_L = 996.3$) as a function of the radius of the bubble for various values of $(\bar{p}_\infty - p_V)$ as indicated.

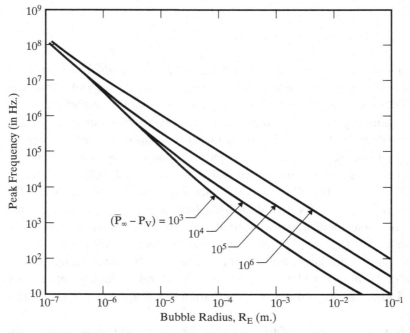

Figure 4.2 Bubble resonant frequency in sodium at $800°K$ ($S = 0.15$, $\mu_L = 0.00023$, $\rho_L = 825.8$) as a function of the radius of the bubble for various values of $(\bar{p}_\infty - p_V)$ as indicated.

wishes to cause cavitation in water by means of an acoustic pressure field, then the frequencies that will be most effective in producing a substantial concentration of large cavitation bubbles will be in this frequency range. This is also the frequency range employed in magnetostrictive devices used to oscillate solid material samples in water (or other liquid) in order to test the susceptibility of that material to cavitation damage (Knapp et al. 1970). Of course, the oscillation of the nuclei produced in this way will be highly nonlinear; nevertheless, the peak response frequency will be less than but not radically different from the peak response frequency for small linear oscillations.

It is also important to note that, like any oscillator, a nucleus excited at its resonant frequency, ω_P, will exhibit a response whose amplitude is primarily a function of the damping. Since the viscous damping is rather small in many practical circumstances, the amplitude given by Equation (4.7) can be very large due to the factor μ_L in the denominator. It could be heuristically argued that this might cause the nucleus to exceed its critical size, R_C (see Section 2.5), and that highly nonlinear behavior with very large amplitudes would result. The pressure amplitude, $\tilde{p}_C$, required to achieve $R_E|\varphi| = R_C - R_E$ can be readily evaluated from Equation (4.7) and the results of the last section:

$$\tilde{p}_C \approx 4\mu_L \left[\omega_N^2 - \frac{4v_L^2}{R_E^4} \right]^{\frac{1}{2}} \left[\left\{ 1 + \frac{\rho_L \omega_N^2 R_E^3}{2S} \right\}^{\frac{1}{2}} - 1 \right] \tag{4.9}$$

and in many circumstances this is approximately equal to $4\mu_L\omega_N$. For a 10 μm nuclei in water at $300°K$ for which the natural frequency is about 10 kHz this critical pressure amplitude is only $0.002\ bar$. Consequently, a nucleus could readily be oscillated in a way that would cause it to exceed the Blake critical radius and therefore proceed to explosive cavitation growth. Of course, nonlinear effects may substantially alter the estimate given in Equation (4.9). Further comment on this and other critical or threshold oscillating pressure levels is delayed until Sections 4.8 and 4.9.

4.3 Effective Polytropic Constant

At this juncture it is appropriate to discuss the validity of the assumption that the gas in the bubble behaves polytropically according to Equation (2.26). For the circumstances of bubble growth and collapse considered in Chapter 2 the polytropic assumption is usually considered acceptable for the following reasons. First, during the growth of a vapor bubble the gas plays a relatively minor role, and the preponderance of vapor will tend to determine the bubble temperature. Second, during the later stages of collapse when the gas predominates, the velocities are so high that an adiabatic assumption, $k = \gamma$, seems appropriate. Since a collapsing bubble loses its spherical symmetry, the resulting internal motions of the gas would, in any case, generate mixing, which would tend to negate any more sophisicated model based on spherical symmetry.

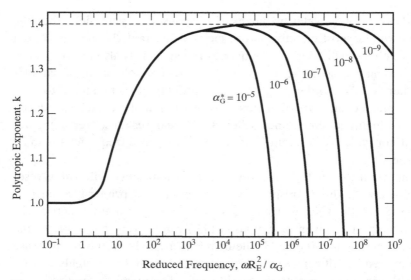

Figure 4.3 Effective polytropic exponent, k, for a diatomic gas ($\gamma = 1.4$) as a function of a reduced frequency, $\omega R_E^2/\alpha_G$, for various values of a reduced thermal diffusivity of the gas, α_G^* (see text). Adapted from Prosperetti (1977b).

The issue of the appropriate polytropic constant is directly coupled with the evaluation of the effective thermal damping of the bubble and was first addressed by Pfriem (1940), Devin (1959), and Chapman and Plesset (1971). Prosperettti (1977b) analysed the problem in detail with particular attention to thermal diffusion in the gas and predicted the effective polytropic exponents shown in Figure 4.3. In that figure the effective polytropic exponent is plotted against a reduced frequency, $\omega R_E^2/\alpha_G$, for various values of a nondimensional thermal diffusivity in the gas, α_G^*, defined by

$$\alpha_G^* = \alpha_G \omega / c_G^2 \tag{4.10}$$

where α_G and c_G are the thermal diffusivity and speed of sound in the gas. Note that for low frequencies (at which there is sufficient time for thermal diffusion) the behavior tends to become isothermal with $k = 1$. On the other hand, at higher frequencies (at which there is insufficient time for heat transfer) the behavior initially tends to become isentropic ($k = \gamma$). At still higher frequencies the mean free path in the gas becomes comparable with the bubble size, and the exponent can take on values outside the range $1 < k < \gamma$ (see Plesset and Prosperetti 1977). Crum (1983) has made measurements of the effective polytropic exponent for bubbles of various gases in water. Figure 4.4 shows typical experimental data for air bubbles in water. The results are consistent with the theory for frequencies below the resonant frequency.

Prosperetti, Crum, and Commander (1988) summarize the current understanding of the theory in which

$$k = \frac{1}{3} Re\{\Upsilon\} \tag{4.11}$$

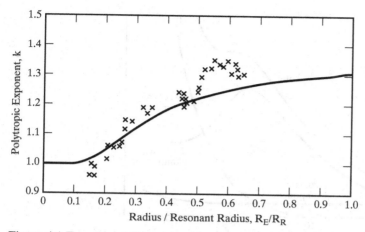

Figure 4.4 Experimentally measured polytropic exponents, k, for air bubbles in water as a function of the bubble radius to resonant bubble radius ratio. The solid line is the theoretical result. Adapted from Crum (1983).

where the complex function, Υ, is given by

$$\Upsilon = \frac{3\gamma}{1 - 3(\gamma - 1)i\chi\left[\left(\frac{i}{\chi}\right)^{\frac{1}{2}} coth\left(\frac{i}{\chi}\right)^{\frac{1}{2}} - 1\right]} \qquad (4.12)$$

where $\chi = \alpha_G/\omega R_E^2$. As we shall discuss in the next section, this analysis also predicts an effective thermal damping that is related to $Im\{\Upsilon\}$.

While the use of an effective polytropic exponent (and the associated thermal damping given by Equation (4.15)) provides a consistent approach for linear oscillations, Prosperetti, Crum, and Commander (1988) have shown that it may cause significant errors when the oscillations become nonlinear. Under these circumstances the behavior of the gas may depart from that which is consistent with an effective polytropic exponent, and there seems to be no option but to numerically solve the detailed mass, momentum, and energy equations in the interior of the bubble.

4.4 Additional Damping Terms

Chapman and Plesset (1971) have presented a useful summary of the three primary contributions to the damping of bubble oscillations, namely that due to liquid viscosity, that due to liquid compressibility through acoustic radiation, and that due to thermal conductivity. It is particularly convenient to represent the three components of damping as three additive contributions to an effective liquid viscosity, μ_E, which can then be employed in the Rayleigh-Plesset equation in place of the actual liquid viscosity, μ_L:

$$\mu_E = \mu_L + \mu_T + \mu_A \qquad (4.13)$$

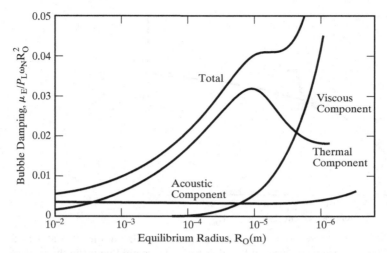

Figure 4.5 Bubble damping components and the total damping as a function of the equilibrium bubble radius, R_E, for water. Damping is plotted as an "effective" viscosity, μ_E, nondimensionalized as shown (from Chapman and Plesset 1971).

where the "acoustic" viscosity, μ_A, is given by

$$\mu_A = \frac{\rho_L \omega^2 R_E^3}{4c_L} \tag{4.14}$$

where c_L is the velocity of sound in the liquid. The "thermal" viscosity, μ_T, follows from the same analysis as was used to obtain the effective polytropic exponent in the preceding section and yields

$$\mu_T = \frac{(\bar{p}_\infty + 2S/R_E)}{4\omega} Im\{\Upsilon\} \tag{4.15}$$

where Υ is given by Equation (4.12).

The relative magnitudes of the three components of damping (or "effective" viscosity) can be quite different for different bubble sizes or radii, R_E. This is illustrated by the data for air bubbles in water at $20°C$ and atmospheric pressure that is taken from Chapman and Plesset (1971) and reproduced as Figure 4.5. Note that the viscous component dominates for very small bubbles, the thermal component is dominant for most bubbles of practical interest, and the acoustic component only dominates for bubbles larger than about 1 *cm*.

4.5 Nonlinear Effects

The preceding sections assume that the perturbation in the bubble radius, φ, is sufficiently small so that the linear approximation holds. However, as Plesset and Prosperetti (1977) have detailed in their review of the subject, single bubbles exhibit a number of interesting and important nonlinear phenomena. When a liquid that will inevitably contain microbubbles is irradiated with sound of a given frequency,

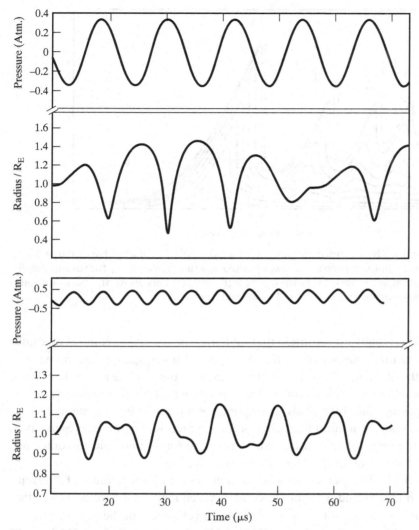

Figure 4.6 Numerically computed examples of the steady nonlinear radial oscillations of a bubble excited by the single-frequency pressure oscillations shown at the top of each graph. Top: Subresonant excitation at 83.4 *kHz* or $\omega/\omega_N = 0.8$ with an amplitude, $\tilde{p} = 0.33$ *bar*. Bottom: Superresonant excitation of a bubble of mean radius 26 *μm* at 191.5 *kHz* or $\omega/\omega_N = 1.8$ with an amplitude $\tilde{p} = 0.33$ *bar*. Adapted from Flynn (1964).

ω, the nonlinear response results in harmonic dispersion, which not only produces harmonics with frequencies that are integer multiples of ω (superharmonics) but, more unusually, subharmonics with frequencies less than ω of the form $m\omega/n$ where m and n are integers. Both the superharmonics and subharmonics become more prominent as the amplitude of excitation is increased. The production of subharmonics was first observed experimentally by Esche (1952), and possible origins of this nonlinear effect were explored in detail by Noltingk and Neppiras (1950, 1951), Flynn (1964), Borotnikova and Soloukin (1964), and Neppiras (1969), among

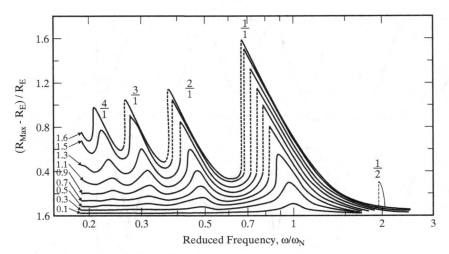

Figure 4.7 Numerically computed amplitudes of radial oscillation of a bubble of radius 1 μm in water at a mean ambient pressure of 1 *bar* plotted as a function of ω/ω_N for various amplitudes of oscillation, $\tilde{p}$ (in *bar*), as shown on the left. The numbers above the peaks indicate the order of the resonance, m/n. Adapted from Lauterborn (1976).

others. Neppiras (1969) also surmised that subharmonic resonance could evolve into transient cavitation. These analytical and numerical investigations use numerical solutions of the Rayleigh-Plesset equation to explore the nonlinear characteristics of a single bubble excited by an oscillating pressure with a single frequency, ω. As might be expected, different kinds of response occur depending on whether ω is greater or less than the natural frequency of the bubble, ω_N. Figure 4.6 presents two examples of the kinds of response encountered, one for $\omega < \omega_N$ and the other for $\omega > \omega_N$. Note the presence of subharmonics in both cases.

Lauterborn (1976) examined numerical solutions for a large number of different excitation frequencies and was able to construct frequency response curves of the kind shown in Figure 4.7. Notice the progressive development of the peak responses at subharmonic frequencies as the amplitude of the excitation is increased. Nonlinear effects not only create these subharmonic peaks but also cause the resonant peaks (both the main resonance near $\omega/\omega_N = 1$ and the subharmonic resonances) to be skewed to the left, creating the discontinuities indicated by the dashed vertical lines. These correspond to bifurcations or sudden transitions between two valid solutions, one with a much larger amplitude than the other. Prosperetti (1977a) has provided a theoretical analysis of these transitions.

4.6 Weakly Nonlinear Analysis

When the amplitudes of oscillation are large, there are no simple analytical methods available, and one must resort to numerical calculations such as those of Lauterborn (1976) in order to investigate the phenomena that result from nonlinearity. However, while the amplitudes are still fairly small, it is valid to use an expansion technique to investigate weakly nonlinear effects. Here we shall retain only terms that are quadratic in the oscillation amplitude; cubic and higher order terms are neglected.

To illustrate weakly nonlinear analysis and the frequency dispersion that results from this procedure, Equations (4.1) and (4.2) are rewritten as

$$P_\infty = \bar{p}_\infty + \sum_{n=1}^{N} Re\left\{\tilde{p}_n e^{jn\delta t}\right\} \tag{4.16}$$

$$R = R_E\left[1 + \sum_{n=1}^{N} Re\left\{\varphi_n e^{jn\delta t}\right\}\right] \tag{4.17}$$

where $n\delta$, $n = 1$ to N, represents a discretization of the frequency domain. When these are substituted into Equation (2.27), all cubic or higher order terms are neglected, and the coefficients of the time-dependent terms are gathered together, the result is the following nonlinear version of Equation (4.3) (Kumar and Brennen 1993):

$$\frac{\tilde{p}_n}{\rho_L \omega_N^2 R_E^2} = \beta_0(n)\varphi_n + \sum_{m=1}^{n-1} \beta_1(n,m)\varphi_m\varphi_{n-m} + \sum_{m=1}^{N-n} \beta_2(n,m)\hat{\varphi}_m\varphi_{n+m} \tag{4.18}$$

where $\hat{\varphi}$ denotes the complex conjugate of φ and

$$\beta_0(n) = \frac{n^2\delta^2}{\omega_N^2} - 1 - j\frac{n\delta}{\omega_N}\frac{4\nu_L}{\omega_N R_E^2} \tag{4.19}$$

$$\beta_1(n,m) = \frac{(3k+1)}{4} + \frac{(3k-1)S}{2\rho_L\omega_N^2 R_E^3} + \frac{\delta^2}{2\omega_N^2}(n-m)\left(n+\frac{m}{2}\right) + j\frac{2\nu_L}{\omega_N R_E^2}\frac{\delta}{\omega_N}(n-m) \tag{4.20}$$

$$\beta_2(n,m) = \frac{(3k+1)}{2} + \frac{(3k-1)S}{\rho_L\omega_N^2 R_E^3} + \frac{\delta^2}{2\omega_N^2}(n^2 - nm - m^2) + j\frac{2\nu_L}{\omega_N R_E^2}\frac{n\delta}{\omega_N} \tag{4.21}$$

Given the fluid and bubble characteristics, Equation (4.18) may be solved iteratively to find φ_n given $\tilde{p}_n$ and the parameters, $\nu/\omega_N R_E^2$, $S/\rho_L\omega_N^2 R_E^3$, k, and δ. The value of N should be large enough to encompass all the harmonics with significant amplitudes.

We shall first examine the characteristics of the radial oscillations that are caused by a single excitation frequency. It is clear from the form of Equation (4.18) that, in this case, the only non-zero φ_n occur at frequencies that are integer multiples of the excitation frequency. Consequently, for this class of problems we may chose δ to be the excitation frequency; then $\tilde{p}_1$ is the amplitude of that excitation and $\tilde{p}_n = 0$ for $n \neq 1$. Figure 4.8 provides two comparisons between the weakly nonlinear solutions and more exact numerical integrations of the Rayleigh-Plesset equation. Clearly these will diverge as the amplitude of oscillation is increased; nevertheless the examples in Figure 4.8 show that the weakly nonlinear solutions are qualitatively valuable.

Figure 4.9 presents examples of the values for $|\varphi_n|$ for three different amplitudes of excitation and demonstrates how the harmonics become more important

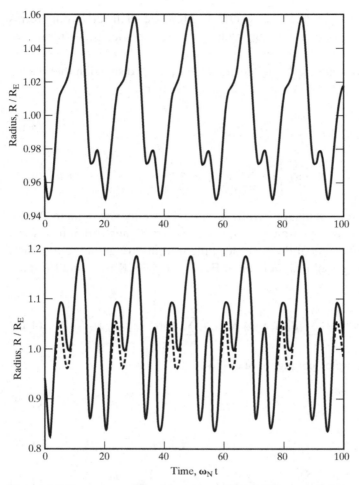

Figure 4.8 Two comparisons between weakly nonlinear solutions and more exact numerical calculations. The parameters are $v_L/\omega_N R_E^2 = 0.01$, $S/\rho_L \omega_N^2 R_E^3 = 0.1$, $k = 1.4$, and the excitation frequency is $\omega_N/3$. The upper figure has a dimensionless excitation amplitude, $\tilde{p}_1/\rho_L \omega_N^2 R_E^2$, of 0.04 while the lower figure has a value of 0.08.

as the amplitude increases. In this example, the frequency of excitation is $\omega_N/6$; the prominence of harmonics close to the natural frequency is characteristic of all solutions in which the excitation frequency is less than ω_N.

Weakly nonlinear solutions can also be used to construct frequency response spectra similar to those due to Lauterborn (1976) described in the preceding section. Figure 4.10 includes examples of such frequency response spectra obtained by plotting the maximum possible deviation from the equilibrium radius, $(R_{max} - R_E)/R_E$, against the excitation frequency. For convenience we estimate $(R_{max} - R_E)/R_E$ as

$$\frac{(R_{max} - R_E)}{R_E} = \sum_{n=1}^{N} |\varphi_n| \tag{4.22}$$

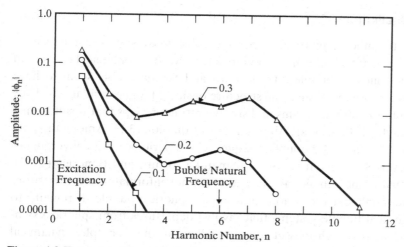

Figure 4.9 Example of the magnitude of the harmonics of radial motion, $|\phi_n|$, from Equation (4.18) with $v_L/\omega_N R_E^2 = 0.01$, $S/\rho_L\omega_N^2 R_E^3 = 0.1$, $k = 1.4$, an excitation frequency of $\omega_N/6$, and three amplitudes of excitation, $\tilde{p}_1/\rho_L\omega_N^2 R_E^2$, as indicated. The connecting lines are for visual effect only.

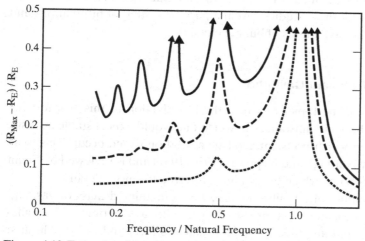

Figure 4.10 Example of frequency response spectra from Equation (4.18) with $v_L/\omega_N R_E^2 = 0.01$, $S/\rho_L\omega_N^2 R_E^3 = 0.1$, $k = 1.4$, and three different excitation amplitudes, $\tilde{p}_1/\rho_L\omega_N^2 R_E^2$, of 0.1 (dotted line), 0.2 (dashed line), and 0.3 (solid line).

Clearly the weakly nonlinear solutions exhibit subharmonic resonances similar to those seen in the more exact solutions like those of Lauterborn (1976). However, they lack some of the finer detail such as the skewing of the resonant peaks that produces the sudden jumps in the response at some subresonant frequencies.

The advantages of the weakly nonlinear analyses become more apparent when dealing with problems of more complex geometry or multiple frequencies of excitation. It is particularly useful in studying the interactions between bubbles in bubble clouds, a subject that is discussed in Chapter 6.

4.7 Chaotic Oscillations

In recent years, the modern methods of nonlinear dynamical systems analysis have led to substantial improvement in the understanding of the nonlinear behavior of bubbles and of clouds of bubbles. Lauterborn and Suchla (1984) seem to have been the first to explore the bifurcation structure of single bubble oscillations. They constructed the bifurcation diagrams and strange attractor maps that result from a compressible Rayleigh-Plesset equation similar to Equation (3.1). Among the phenomena obtained was a period doubling sequence of a periodic orbit converging to a strange attractor. Subsequent studies by Smereka, Birnir, and Banerjee (1987), Parlitz et al. (1990), and others have provided further information on the nature of these chaotic, nonlinear oscillations of a single, spherical bubble. It remains to be seen how far real bubble systems that involve departures from spherical symmetry and from the Rayleigh-Plesset equation adhere to these complex dynamical behaviors.

In Section 6.10, we shall explore the linear, dynamic behavior of a cloud of bubbles and will find that such clouds exhibit their own characteristic dynamics and natural frequencies. The nonlinear, chaotic behavior of clouds of bubbles have also been recently examined by Smereka and Banerjee (1988) and Birnir and Smereka (1990), and these studies reveal a parallel system of bifurcations and strange attractors in the oscillations of bubble clouds.

4.8 Threshold for Transient Cavitation

We now turn to one of the topics raised in the introduction to this chapter: the distinction between those circumstances in which one would expect stable acoustic cavitation and those in which transient acoustic cavitation would occur. This issue was first addressed by Noltingk and Neppiras (1950, 1951) and is reviewed by Flynn (1964) and Young (1989), to which the reader is referred for more detail.

We consider a bubble of equilibrium size, R_E, containing a mass of gas, m_G, and subjected to a mean ambient pressure, $\bar{p}_\infty$, with a superimposed oscillation of frequency, ω, and amplitude, $\tilde{p}$ (see Equations (4.1) and (4.2)). The first step in establishing the criterion is accomplished by the static stability analysis of Section 2.5. There we explored the stability of a bubble when the pressure far from the bubble was varied and identified a critical size, R_C, and a critical threshold pressure, $p_{\infty c}$, which, if reached, would lead to unstable bubble growth and therefore, in the present context, to transient cavitation. The added complication here is that there is only a finite time during each cycle during which growth can occur, so one must address the issue of whether or not that time is sufficient for significant unstable growth.

The issue is determined by the relationship between the radian frequency, ω, of the imposed oscillations and the natural frequency, ω_N, of the bubble. If $\omega \ll \omega_N$, then the liquid inertia is relatively unimportant in the bubble dynamics and the bubble will respond quasistatically. Under these circumstances the Blake criterion (see

Section 2.5) will hold. Denoting the critical amplitude at which transient cavitation will occur by $\tilde{p}_C$, it follows that the critical conditions will be reached when the minimum instantaneous pressure, $(\bar{p}_\infty - \tilde{p})$, just reaches the critical Blake threshold pressure given by Equation (2.45). Therefore

$$\tilde{p}_C = \bar{p}_\infty - p_V + \frac{4S}{3}\left[\frac{8\pi S}{9m_G T_B K_G}\right]^{\frac{1}{2}} \tag{4.23}$$

On the other hand, if $\omega \gg \omega_N$, the issue will involve the dynamics of bubble growth since inertia will determine the size of the bubble perturbations. The details of this bubble dynamic problem have been addressed by Flynn (1964) and convenient guidelines are provided by Apfel (1981). Following Apfel's construction, we note that a neccessary but not sufficient condition for transient cavitation is that the ambient pressure, p_∞, fall below the vapor pressure for part of the oscillation cycle. The typical negative pressure will, of course, be given by $(\bar{p}_\infty - \tilde{p})$. Moreover, the pressure will be negative for some fraction of the period of oscillation; that fraction is solely related to the parameter, $\beta = (1 - \bar{p}_\infty/\tilde{p})$ (Apfel 1981). Then, assuming that the quasistatic Blake threshold has been exceeded, the bubble growth rate will be given roughly by the asymptotic growth rate of Equation (2.33). Combining this with the time available for growth, the typical maximum bubble radius, R_M, will be given by

$$R_M = f(\beta)\frac{\pi}{\omega}\left[\frac{\tilde{p}-\bar{p}_\infty}{\rho_L}\right]^{\frac{1}{2}} \tag{4.24}$$

where we have neglected the vapor pressure, p_V. In this expression the function $f(\beta)$ accounts for some of the details such as the fraction of the half-period, π/ω, for which the pressure is negative. Apfel (1981) finds

$$f(\beta) = \left(\frac{4}{3\pi}\right)(2\beta)^{\frac{1}{2}}\left\{1 + \frac{2}{3(1-\beta)}\right\}^{\frac{1}{3}} \tag{4.25}$$

The final step in constructing the criterion for $\omega \gg \omega_N$ is to argue that transient cavitation will occur when $R_M \to 2R_E$ and, using this, the critical pressure becomes

$$\tilde{p}_C = \bar{p}_\infty + 4\rho_L R_E^2 \omega^2/\pi^2 f^2 \tag{4.26}$$

For more detailed analyses the reader is referred to the work of Flynn (1964) and Apfel (1981).

4.9 Rectified Mass Diffusion

We now shift attention to a different nonlinear effect involving the mass transfer of dissolved gas between the liquid and the bubble. This important nonlinear diffusion effect occurs in the presence of an acoustic field and is known as "rectified mass diffusion" (Blake 1949a). Analytical models of this phenomenon were first put forward by Hsieh and Plesset (1961) and Eller and Flynn (1965), and reviews of the subject can be found in Crum (1980, 1984) and Young (1989).

Consider a gas bubble in a liquid with dissolved gas as described in Section 2.6. Now, however, we add an oscillation to the ambient pressure. Gas will tend to come out of solution into the bubble during that part of the oscillation cycle when the bubble is larger than the mean because the partial pressure of gas in the bubble is then depressed. Conversely, gas will redissolve during the other half of the cycle when the bubble is smaller than the mean. The linear contributions to the mass of gas in the bubble will, of course, balance so that the average gas content in the bubble will not be affected at this level. However, there are two nonlinear effects that tend to increase the mass of gas in the bubble. The first of these is due to the fact that release of gas by the liquid occurs during that part of the cycle when the surface area is larger, and therefore the influx during that part of the cycle is slightly larger than the efflux during the part of the cycle when the bubble is smaller. Consequently, there is a net flux of gas into the bubble which is quadratic in the perturbation amplitude. Second, the diffusion boundary layer in the liquid tends to be stretched thinner when the bubble is larger, and this also enhances the flux into the bubble during the part of the cycle when the bubble is larger. This effect contributes a second, quadratic term to the net flux of gas into the bubble. Recent analyses, which include all of the contributing nonlinear terms (see Crum 1984 or Young 1989), yield the following modification to the steady mass diffusion result given previously in Equation (2.56) (see Section 2.6):

$$R_E \frac{dR_E}{dt} = \frac{D}{\rho_{GE}} \left[\frac{(c_\infty - c_S \Gamma_3 (1 + 2S/R_E \bar{p}_\infty)/\Gamma_2)}{(1 + 4S/3R_E \bar{p}_\infty)} \right] \left[\Gamma_1 + \frac{R_E (\Gamma_2)^{\frac{1}{2}}}{(\pi D t)^{\frac{1}{2}}} \right] \tag{4.27}$$

which is identical with Equation (2.56) except for the Γ terms, which differ from unity by terms that are quadratic in the fluctuating pressure amplitude, $\tilde{p}$:

$$\Gamma_1 = 1 + \Theta_1 \Theta_2^{-2} \left(\frac{\tilde{p}}{\bar{p}_\infty} \right)^2 \tag{4.28}$$

$$\Gamma_2 = 1 + (4\Theta_1 + 3)\Theta_2^{-2} \left(\frac{\tilde{p}}{\bar{p}_\infty} \right)^2 \tag{4.29}$$

$$\Gamma_3 = 1 + (4 - 3k)(\Theta_1 - 3(k-1)/4)\Theta_2^{-2} \left(\frac{\tilde{p}}{\bar{p}_\infty} \right)^2 \tag{4.30}$$

where

$$\Theta_1 = \frac{(3k + 1 - \beta^2)/4 + (S/4R_E \bar{p}_\infty)(6k + 2 - 4/3k)}{1 + (2S/R_E \bar{p}_\infty)(1 - 1/3k)} \tag{4.31}$$

$$\Theta_2 = \left(\frac{\rho_L R_E^2}{\bar{p}_\infty} \right) \left[(\omega^2 - \omega_N^2)^2 + (4\mu\omega^2 \omega_N / 3k\bar{p}_\infty)^2 \right]^{\frac{1}{2}} \tag{4.32}$$

$$\beta^2 = \frac{\rho_L \omega^2 R_E^2}{3k\bar{p}_\infty} \tag{4.33}$$

where one must choose an appropriate μ to represent the total effective damping (see Section 4.4) and an appropriate effective polytropic constant, k (see

Section 4.3). Valuable contributions to the evolution of these results were made by Hsieh and Plesset (1961), Eller and Flynn (1965), Safar (1968), Eller (1969, 1972, 1975), Skinner (1970), and Crum (1980, 1984), among others.

Strasberg (1961) first explored the issue of the conditions under which a bubble would grow due to rectified diffusion. Clearly, the sign of the bubble growth rate predicted by Equation (4.27) will be determined by the sign of the term

$$c_\infty - c_S(1 + 2S/R_E\bar{p}_\infty)\left(\frac{\Gamma_3}{\Gamma_2}\right) \tag{4.34}$$

In the absence of oscillations and surface tension, this leads to the conclusion that the bubble will grow when $c_\infty > c_S$ and will dissolve when the reverse is true. The term involving surface tension causes bubbles in a saturated solution ($c_S = c_\infty$) to dissolve but usually has only a minor effect in real applications. However, in the presence of oscillations the term Γ_3/Γ_2 will decrease below unity as the amplitude, $\tilde{p}$, is increased. This causes a positive increment in the growth rate as anticipated earlier. Even in a subsaturated liquid for which $c_\infty < c_S$ this increment could cause the sign of dR/dt to change and become positive. Thus Equation (4.27) allows us to quantify the bubble growth rate due to rectified mass diffusion.

If an oscillating pressure is applied to a fluid consisting of a subsaturated or saturated liquid and seeded with microbubbles of radius, R_E, then Expression (4.34) also demonstates that there will exist a certain critical or threshold amplitude above which the microbubbles will begin to grow by rectified diffusion. This threshold amplitude, $\tilde{p}_C$, will be large enough so that the value of Γ_3/Γ_2 is sufficiently small to make Expression (4.34) vanish. From Equations (4.29) to (4.33) the threshold amplitude becomes

$$\tilde{p}_C^2 = \frac{(\rho_L R_E^2 \omega_N^2)^2\left[\left(1 - \frac{\omega^2}{\omega_N^2}\right)^2 + \left(\frac{4\mu_L\omega^2}{3k\omega N\bar{p}_\infty}\right)^2\right]\left[1 + \frac{2S}{R_E\bar{p}_\infty} - \frac{c_\infty}{c_S}\right]}{(3 + 4\Theta_1)\left(\frac{c_\infty}{c_S}\right) - \left[\frac{3(k-1)(3k-4)}{4} + (4 - 3k)\Theta_1\right]\left[1 + \frac{2S}{R_E\bar{p}_\infty}\right]} \tag{4.35}$$

Typical experimental measurements of the rates of growth and of the threshold pressure amplitudes are shown in Figures 4.11 and 4.12. The data are from the work of Crum (1980, 1984) and are for distilled water that is saturated with air. It is clear that there is satisfactory agreement for the cases shown. However, Crum also observed significant discrepancies when a surface-active agent was added to the water to change the surface tension.

Finally, we note again that most of the theories assume spherical symmetry and that departure from sphericity could alter the diffusion boundary layer in ways that could radically affect the mass transfer process. Furthermore, there is some evidence that acoustic streaming induced by the excitation can also cause disruption of the diffusion boundary layer (Elder 1959, Gould 1966).

Before leaving the subject of rectified diffusion, it is important to emphasize that the bubble growth that it causes is very slow compared with most of the other growth processes considered in the last two chapters. It is appropriate to think of

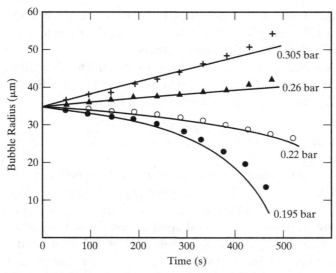

Figure 4.11 Examples from Crum (1980) of the growth (or shrinkage) of air bubbles in saturated water ($S = 68$ *dynes/cm*) due to rectified diffusion. Data is shown for four pressure amplitudes as shown. The lines are the corresponding theoretical predictions.

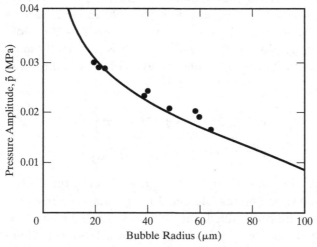

Figure 4.12 Data from Crum (1984) of the threshold pressure amplitude for rectified diffusion for bubbles in distilled water ($S = 68$ *dynes/cm*) saturated with air. The frequency of the sound is 22.1 *kHz*. The line is the prediction of Equation (4.35).

it as causing a gradual, quasistatic change in the equilibrium size of the bubble, R_E. However, it does provide a mechanism by which very small and stable nuclei might grow sufficiently to become nuclei for cavitation. It is also valuable to observe that the Blake threshold pressure, $p_{\infty c}$, increases as the mass of gas in the bubble, m_G, increases (see Equation (2.46)). Therefore, as m_G increases, a smaller reduction in the pressure is necessary to create an unstable bubble. That is to say, it becomes easier to cavitate the liquid.

4.10 Bjerknes Forces

A different nonlinear effect is the force experienced by a bubble in an acoustic field due to the finite wavelength of the sound waves. The spatial wavenumber will be denoted by $k = \omega/c_L$. The presence of such waves implies an instantaneous pressure gradient in the liquid. To model this we substitute

$$\tilde{p} = \tilde{p}^* \sin(kx_i) \tag{4.36}$$

into Equation (4.1) where the constant $\tilde{p}^*$ is the amplitude of the sound waves and x_i is the direction of wave propagation. Like any other pressure gradient, this produces an instantaneous force, F_i, on the bubble in the x_i direction given by

$$F_i = -\frac{4}{3}\pi R^3 \left(\frac{dp_\infty}{dx_i}\right) \tag{4.37}$$

Since both R and dp_∞/dx_i contain oscillating components, it follows that the combination of these in Equation (4.37) will lead to a nonlinear, time-averaged component in F_i, which we will denote by $\bar{F}_i$. Substituting Equations (4.36), (4.1), and (4.2) into (4.37) yields

$$\bar{F}_i = -2\pi R_E^3 Re\{\varphi\}k\tilde{p}^* \cos(kx_i) \tag{4.38}$$

where the radial oscillation amplitude, φ, is given by Equation (4.3) so that

$$Re\{\varphi\} = \frac{\tilde{p}(\omega^2 - \omega_N^2)}{\rho_L R_E^2 \left[(\omega^2 - \omega_N^2)^2 + \left(4\nu_L\omega/R_E^2\right)^2\right]} \tag{4.39}$$

If ω is not too close to ω_N, a useful approximation is

$$Re\{\varphi\} \approx \tilde{p}/\rho_L R_E^2(\omega^2 - \omega_N^2) \tag{4.40}$$

and substituting this into Equation (4.38) yields

$$\bar{F}_i = -\frac{\pi k R_E(\tilde{p}^*)^2 \sin(2kx_i)}{\rho_L(\omega^2 - \omega_N^2)} \tag{4.41}$$

This is known as the primary Bjerknes force since it follows from some of the effects discussed by that author (Bjerknes 1909). The effect was first properly identified by Blake (1949b).

The form of the primary Bjerknes force produces some interesting bubble migration patterns in a stationary sound field. Note from Equation (4.41) that if the excitation frequency, ω, is less than the natural frequency, ω_N, (or $R_E < R_R$) then the primary Bjerknes force will cause migration of the bubbles away from the nodes in the pressure field and toward the antinodes (points of largest pressure amplitude). On the other hand, if $\omega > \omega_N$ (or $R_E > R_R$) the bubbles will tend to migrate from the antinodes to the nodes. A number of investigators (for example, Crum and Eller 1970) have observed the process by which small bubbles in a stationary sound field first migrate to the antinodes, where they grow by rectified diffusion until they are larger than the resonant radius. They then migrate back to the nodes, where they may dissolve again when they experience only small pressure oscillations. Crum and Eller (1970) and have shown that the translational velocities of migrating bubbles are compatible with the Bjerknes force estimates given above.

Finally, it is important to mention one other nonlinear effect. An acoustic field can cause time-averaged or mean motions in the fluid itself. These are referred to as *acoustic streaming*,. The term *microstreaming* is used to refer to such motions near a small bubble. Generally these motions take the form of circulation patterns and, in a classic paper, Elder (1959) observed and recorded the circulating patterns of microstreaming near the surface of small gas bubbles in liquids. As stated earlier, these circulation patterns could alter the processes of heat and mass diffusion to or from a bubble and therefore modify phenomena such as rectified diffusion.

References

Apfel, R.E. (1981). Acoustic cavitation prediction. *J. Acoust. Soc. Am.*, **69**, 1624–1633.

Birnir, B. and Smereka, P. (1990). Existence theory and invariant manifolds of bubble clouds. *Comm. Pure Appl. Math.*, **43**, 363–413.

Bjerknes, V. (1909). *Die Kraftfelder*. Friedrich Vieweg and Sohn, Braunsweig.

Blake, F.G. (1949a). The onset of cavitation in liquids. *Acoustics Res. Lab., Harvard Univ., Tech. Memo. No. 12*.

Blake, F.G. (1949b). Bjerknes forces in stationary sound fields. *J. Acoust. Soc. Am.*, **21**, 551.

Borotnikova, M.I. and Soloukin, R.I. (1964). A calculation of the pulsations of gas bubbles in an incompressible liquid subject to a periodically varying pressure. *Sov. Phys. Acoust.*, **10**, 28–32.

Chapman, R.B. and Plesset, M.S. (1971). Thermal effects in the free oscillation of gas bubbles. *ASME J. Basic Eng.*, **93**, 373–376.

Cole, R.H. (1948). *Underwater explosions*. Princeton Univ. Press, reprinted by Dover in 1965.

Crum, L.A. (1979). Tensile strength of water. *Nature*, **278**, 148–149.

Crum, L.A. (1980). Measurements of the growth of air bubbles by rectified diffusion. *J. Acoust. Soc. Am.*, **68**, 203–211.

Crum, L.A. (1983). The polytropic exponent of gas contained within air bubbles pulsating in a liquid. *J. Acoust. Soc. Am.*, **73**, 116–120.

Crum, L.A. (1984). Rectified diffusion. *Ultrasonics*, **22**, 215–223.

Crum, L.A. and Eller, A.I. (1970). Motion of bubbles in a stationary sound field. *J. Acoust. Soc. Am.*, **48**, 181–189.

Devin, C. (1959). Survey of thermal, radiation, and viscous damping of pulsating air bubbles in water. *J. Acoust. Soc. Am.*, **31**, 1654–1667.

Elder, S.A. (1959). Cavitation microstreaming. *J. Acoust. Soc. Am.*, **31**, 54–64.

Eller, A.I. and Flynn, H.G. (1965). Rectified diffusion during non-linear pulsation of cavitation bubbles. *J. Acoust. Soc. Am.*, **37**, 493–503.

Eller, A.I. (1969). Growth of bubbles by rectified diffusion. *J. Acoust. Soc. Am.*, **46**, 1246–1250.

Eller, A.I. (1972). Bubble growth by diffusion in an 11 *kHz* sound field. *J. Acoust. Soc. Am.*, **52**, 1447–1449.

Eller, A.I. (1975). Effects of diffusion on gaseous cavitation bubbles. *J. Acoust. Soc. Am.*, **57**, 1374–1378.

Esche, R. (1952). Untersuchung der Schwingungskavitation in Flüssigkeiten. *Acustica*, **2**, AB208–AB218.

Flynn, H.G. (1964). Physics of acoustic cavitation in liquids. *Physical Acoustics*, **1B**. Academic Press.

Gould, R.K. (1966). Heat transfer across a solid-liquid interface in the presence of acoustic streaming. *J. Acoust. Soc. Am.*, **40**, 219–225.

Hsieh, D.-Y. and Plesset, M.S. (1961). Theory of rectified diffusion of mass into gas bubbles. *J. Acoust. Soc. Am.*, **33**, 206–215.

Knapp, R.T., Daily, J.W., and Hammitt, F.G. (1970). *Cavitation*. McGraw-Hill, New York.

Kumar, S. and Brennen, C.E. (1993). Some nonlinear interactive effects in bubbly cavitation clouds. *J. Fluid Mech.*, **253**, 565–591.

Lauterborn, W. (1976). Numerical investigation of nonlinear oscillations of gas bubbles in liquids. *J. Acoust. Soc. Am.*, **59**, 283–293.

Lauterborn, W. and Suchla, E. (1984). Bifurcation superstructure in a model of acoustic turbulence. *Phys. Rev. Lett.*, **53**, 2304–2307.

Neppiras, E.A. (1969). Subharmonic and other low-frequency emission from bubbles in sound-irradiated liquids. *J. Acoust. Soc. Am.*, **46**, 587–601.

Neppiras, E.A. (1980). Acoustic cavitation. *Phys. Rep.*, **61**, 160–251.

Neppiras, E.A. and Noltingk, B.E. (1951). Cavitation produced by ultrasonics: theoretical conditions for the onset of cavitation. *Proc. Phys. Soc., London*, **64B**, 1032–1038.

Noltingk, B.E. and Neppiras, E.A. (1950). Cavitation produced by ultrasonics. *Proc. Phys. Soc., London*, **63B**, 674–685.

Parlitz, U., Englisch, V., Scheffczyk, C., and Lauterborn, W. (1990). Bifurcation structure of bubble oscillators. *J. Acoust. Soc. Am.*, **88**, 1061–1077.

Pfriem, H. (1940). *Akust. Zh.*, **5**, 202–212.

Plesset, M.S. and Prosperetti, A. (1977). Bubble dynamics and cavitation. *Ann. Rev. Fluid Mech.*, **9**, 145–185.

Prosperetti, A. (1974). Nonlinear oscillations of gas bubbles in liquids: steady state solutions. *J. Acoust. Soc. Am.*, **56**, 878–885.

Prosperetti, A. (1977a). Application of the subharmonic threshold to the measurement of the damping of oscillating gas bubbles. *J. Acoust. Soc. Am.*, **61**, 11–16.

Prosperetti, A. (1977b). Thermal effects and damping mechanisms in the forced radial oscillations of gas bubbles in liquids. *J. Acoust. Soc. Am.*, **61**, 17–27.

Prosperetti, A. (1982). Bubble dynamics: a review and some recent results. *Appl. Sci. Res.*, **38**, 145–164.

Prosperetti, A. (1984). Bubble phenomena in sound fields: part one and part two. *Ultrasonics*, **22**, 69–77 and 115–124.

Prosperetti, A., Crum, L.A. and Commander, K.W. (1988). Nonlinear bubble dynamics. *J. Acoust. Soc. Am.*, **83**, 502–514.

Safar, M.H. (1968). Comments on papers concerning rectified diffusion of cavitation bubbles. *J. Acoust. Soc. Am.*, **43**, 1188–1189.

Skinner, L.A. (1970). Pressure threshold for acoustic cavitation. *J. Acoust. Soc. Am.*, **47**, 327–331.

Smereka, P., Birnir, B. and Banerjee, S. (1987). Regular and chaotic bubble oscillations in periodically driven pressure fields. *Phys. Fluids*, **30**, 3342–3350.

Smereka, P. and Banerjee, S. (1988). The dynamics of periodically driven bubble clouds. *Phys. Fluids*, **31**, 3519–3531.

Strasberg, M. (1961). Rectified diffusion: comments on a paper of Hsieh and Plesset. *J. Acoust. Soc. Am.*, **33**, 359.

Young, F.R. (1989). *Cavitation*. McGraw-Hill Book Company.

5 Translation of Bubbles

5.1 Introduction

This chapter will briefly review the issues and problems involved in constructing the equations of motion for individual bubbles (or drops or solid particles) moving through a fluid and will therefore focus on the dynamics of relative motion rather than the dynamics of growth and collapse. For convenience we shall use the generic name "particle" when any or all of bubbles, drops, and solid particles are being considered. The analyses are implicitly confined to those circumstances in which the interactions between neighboring particles are negligible. In very dilute multiphase flows in which the particles are very small compared with the global dimensions of the flow and are very far apart compared with the particle size, it is often sufficient to solve for the velocity and pressure, $u_i(x_i, t)$ and $p(x_i, t)$, of the continuous suspending fluid while ignoring the particles or disperse phase. Given this solution one could then solve an equation of motion for the particle to determine its trajectory. This chapter will focus on the construction of such a particle or bubble equation of motion. Interactions between particles or, more particularly, bubble, are left for later.

The body of fluid mechanical literature on the subject of flows around particles or bodies is very large indeed. Here we present a summary that focuses on a spherical particle of radius, R, and employs the following common notation. The components of the translational velocity of the center of the particle will be denoted by $V_i(t)$. The velocity that the fluid would have had at the location of the particle center in the absence of the particle will be denoted by $U_i(t)$. Note that such a concept is difficult to extend to the case of interactive multiphase flows. Finally, the velocity of the particle relative to the fluid is denoted by $W_i(t) = V_i - U_i$.

Frequently the approach used to construct equations for $V_i(t)$ (or $W_i(t)$) given $U_i(x_i, t)$ is to individually estimate all the fluid forces acting on the particle and to equate the total fluid force, F_i, to $m_p dV_i/dt$ (where m_p is the particle mass, assumed constant). These fluid forces may include forces due to buoyancy, added mass, drag, etc. In the absence of fluid acceleration ($dU_i/dt = 0$) such an approach can be made unambiguously; however, in the presence of fluid acceleration, this kind of heuristic approach can be misleading. Hence we concentrate in the next few sections on a fundamental fluid mechanical approach, which minimizes possible ambiguities. The classical results for a spherical particle or bubble are reviewed first. The analysis is confined to a suspending fluid that is incompressible and Newtonian so that the basic equations to be solved are the continuity equation

$$\frac{\partial u_j}{\partial x_j} = 0 \tag{5.1}$$

and the Navier-Stokes equations

$$\rho \left\{ u_j \frac{\partial u_i}{\partial x_j} \right\} = -\frac{\partial p}{\partial x_i} + \rho v \frac{\partial^2 u_i}{\partial x_j \partial x_j} \tag{5.2}$$

where ρ and v are the density and kinematic viscosity of the suspending fluid. It is assumed that the only external force is that due to gravity, g. Then the actual pressure is $p' = p - \rho g z$ where z is a coordinate measured vertically upward.

Furthermore, in order to maintain clarity we confine attention to rectilinear relative motion in a direction conveniently chosen to be the x_1 direction.

5.2 High Re Flows Around a Sphere

For steady flows about a sphere in which $dU_i/dt = dV_i/dt = dW_i/dt = 0$, it is convenient to use a coordinate system, x_i, fixed in the particle as well as polar coordinates (r, θ) and velocities u_r, u_θ as defined in Figure 5.1.

Then Equations (5.1) and (5.2) become

$$\frac{1}{r^2}\frac{\partial}{\partial r}(r^2 u_r) + \frac{1}{r\sin\theta}\frac{\partial}{\partial\theta}(u_\theta \sin\theta) = 0 \tag{5.3}$$

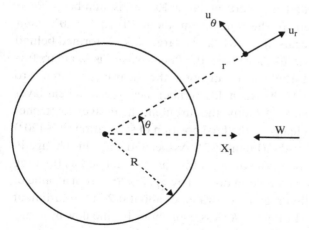

Figure 5.1 Notation for a spherical particle.

and

$$\rho \left\{ \frac{\partial u_r}{\partial t} + u_r \frac{\partial u_r}{\partial r} + \frac{u_\theta}{r} \frac{\partial u_r}{\partial \theta} - \frac{u_\theta^2}{r} \right\} = -\frac{\partial p}{\partial r}$$

$$+ \rho \nu \left\{ \frac{1}{r^2} \frac{\partial}{\partial r} \left(r^2 \frac{\partial u_r}{\partial r} \right) + \frac{1}{r^2 \sin\theta} \frac{\partial}{\partial \theta} \left(\sin\theta \frac{\partial u_r}{\partial \theta} \right) - \frac{2 u_r}{r^2} - \frac{2}{r^2} \frac{\partial u_\theta}{\partial \theta} \right\} \tag{5.4}$$

$$\rho \left\{ \frac{\partial u_\theta}{\partial t} + u_r \frac{\partial u_\theta}{\partial r} + \frac{u_\theta}{r} \frac{\partial u_\theta}{\partial \theta} + \frac{u_r u_\theta}{r} \right\} = -\frac{1}{r} \frac{\partial p}{\partial \theta}$$

$$+ \rho \nu \left\{ \frac{1}{r^2} \frac{\partial}{\partial r} \left(r^2 \frac{\partial u_\theta}{\partial r} \right) + \frac{1}{r^2 \sin\theta} \frac{\partial}{\partial \theta} \left(\sin\theta \frac{\partial u_\theta}{\partial \theta} \right) + \frac{2}{r^2} \frac{\partial u_r}{\partial \theta} - \frac{u_\theta}{r^2 \sin^2\theta} \right\} \tag{5.5}$$

The Stokes streamfunction, ψ, is defined to satisfy continuity automatically:

$$u_r = \frac{1}{r^2 \sin\theta} \frac{\partial \psi}{\partial \theta}; \quad u_\theta = -\frac{1}{r \sin\theta} \frac{\partial \psi}{\partial r} \tag{5.6}$$

and the inviscid potential flow solution is

$$\psi = -\frac{W r^2}{2} \sin^2\theta - \frac{D}{r} \sin^2\theta \tag{5.7}$$

$$u_r = -W \cos\theta - \frac{2D}{r^3} \cos\theta \tag{5.8}$$

$$u_\theta = +W \sin\theta - \frac{D}{r^3} \sin\theta \tag{5.9}$$

$$\phi = -W r \cos\theta + \frac{D}{r^2} \cos\theta \tag{5.10}$$

where, because of the boundary condition $(u_r)_{r=R} = 0$, it follows that $D = -WR^3/2$. In potential flow one may also define a velocity potential, ϕ, such that $u_i = \partial\phi/\partial x_i$. The classic problem with such solutions is the fact that the drag is zero, a circumstance termed D'Alembert's paradox. The flow is symmetric about the $x_2 x_3$ plane through the origin and there is no wake.

The real viscous flows around a sphere at large Reynolds numbers, $Re = 2WR/\nu > 1$, are well documented. In the range from about 10^3 to 3×10^5, laminar boundary layer separation occurs at $\theta \cong 84°$ and a large wake is formed behind the sphere (see Figure 5.2). Close to the sphere the "near-wake" is laminar; further downstream transition and turbulence occurring in the shear layers spreads to generate a turbulent "far-wake." As the Reynolds number increases the shear layer transition moves forward until, quite abruptly, the turbulent shear layer reattaches to the body, resulting in a major change in the final position of separation ($\theta \cong 120°$) and in the form of the turbulent wake (Figure 5.2). Associated with this change in flow pattern is a dramatic decrease in the drag coefficient, C_D (defined as the drag force on the body in the negative x_1 direction divided by $\frac{1}{2}\rho W^2 \pi R^2$), from a value of about 0.5 in the laminar separation regime to a value of about 0.2 in the turbulent separation regime (Figure 5.3). At values of Re less than about 10^3 the flow becomes quite unsteady with periodic shedding of vortices from the sphere.

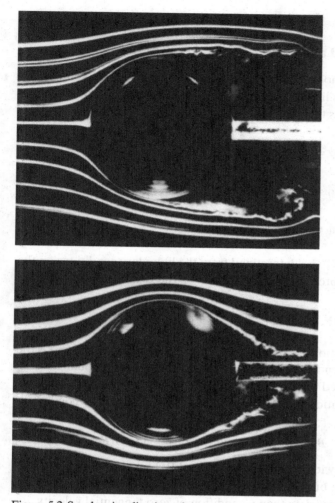

Figure 5.2 Smoke visualization of the nominally steady flows (from left to right) past a sphere showing, at the top, laminar separation at $Re = 2.8 \times 10^5$ and, on the bottom, turbulent separation at $Re = 3.9 \times 10^5$. Photographs by F.N.M. Brown, reproduced with the permission of the University of Notre Dame.

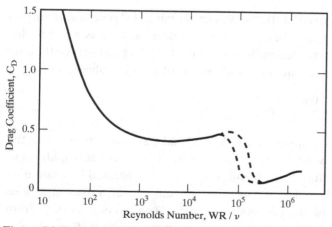

Figure 5.3 Drag coefficient on a sphere as a function of Reynolds number. Dashed curves indicate the drag crisis regime in which the drag is very sensitive to other factors such as the free stream turbulence.

5.3 Low Re Flows Around a Sphere

At the other end of the Reynolds number spectrum is the classic Stokes solution for flow around a sphere. In this limit the terms on the left-hand side of Equation (5.2) are neglected and the viscous term retained. This solution has the form

$$\psi = \sin^2\theta \left[-\frac{Wr^2}{2} + \frac{A}{r} + Br \right] \tag{5.11}$$

$$u_r = \cos\theta \left[-W + \frac{2A}{r^3} + \frac{2B}{r} \right] \tag{5.12}$$

$$u_\theta = -\sin\theta \left[-W - \frac{A}{r^3} + \frac{B}{r} \right] \tag{5.13}$$

where A and B are constants to be determined from the boundary conditions on the surface of the sphere. The force, F, on the "particle" in the x_1 direction is

$$F_1 = \frac{4}{3}\pi R^2 \rho \nu \left[-\frac{4W}{R} + \frac{8A}{R^4} + \frac{2B}{R^2} \right] \tag{5.14}$$

Several subcases of this solution are of interest in the present context. The first is the classic Stokes (1851) solution for a solid sphere in which the no-slip boundary condition, $(u_\theta)_{r=R} = 0$, is applied (in addition to the kinematic condition $(u_r)_{r=R} = 0$). This set of boundary conditions, referred to as the Stokes boundary conditions, leads to

$$A = -\frac{WR^3}{4}, \quad B = +\frac{3WR}{4} \quad \text{and} \quad F_1 = -6\pi\rho\nu WR \tag{5.15}$$

The second case originates with Hadamard (1911) and Rybczynski (1911) who suggested that, in the case of a bubble, a condition of zero shear stress on the sphere surface would be more appropriate than a condition of zero tangential velocity, u_θ. Then it transpires that

$$A = 0, \quad B = +\frac{WR}{2} \quad \text{and} \quad F_1 = -4\pi\rho\nu WR \tag{5.16}$$

Real bubbles may conform to either the Stokes or Hadamard-Rybczynski solutions depending on the degree of contamination of the bubble surface, as we shall discuss in more detail in the next section. Finally, it is of interest to observe that the potential flow solution given in Equations (5.7) to (5.10) is also a subcase with

$$A = +\frac{WR^3}{2}, \quad B = 0 \quad \text{and} \quad F_1 = 0 \tag{5.17}$$

However, another paradox, known as the Whitehead paradox, arises when the validity of these Stokes flow solutions at small (rather than zero) Reynolds numbers is considered. The nature of this paradox can be demonstrated by examining the magnitude of the neglected term, $u_j \partial u_i / \partial x_j$, in the Navier-Stokes equations relative to the magnitude of the retained term $\nu \partial^2 u_i / \partial x_j \partial x_j$. As is evident from Equation (5.11), far from the sphere the former is proportional to $W^2 R / r^2$ whereas the latter behaves like $\nu WR/r^3$. It follows that although the retained term will dominate close to the body (provided the Reynolds number $Re = 2WR/\nu \ll 1$), there will always be a radial position, r_c, given by $R/r_c = Re$ beyond which the neglected term will exceed the retained viscous term. Hence, even if $Re \ll 1$, the Stokes solution is not uniformly valid. Recognizing this limitation, Oseen (1910) attempted to correct the Stokes solution by retaining in the basic equation an approximation to $u_j \partial u_i/\partial x_j$ that would be valid in the far field, $-W\partial u_i/\partial x_1$. Thus the Navier-Stokes equations are approximated by

$$-W\frac{\partial u_i}{\partial x_1} = -\frac{1}{\rho}\frac{\partial p}{\partial x_i} + \nu\frac{\partial^2 u_i}{\partial x_j \partial x_j} \tag{5.18}$$

Oseen was able to find a closed form solution to this equation that satisfies the Stokes boundary conditions approximately:

$$\psi = -WR^2 \left[\frac{r^2 \sin^2\theta}{2R^2} + \frac{R\sin^2\theta}{4r} + \frac{3\nu(1+\cos\theta)}{2WR}\left\{ 1 - e^{\frac{Wr}{2\nu(1-\cos\theta)}} \right\} \right] \tag{5.19}$$

which yields a drag force

$$F_1 = -6\pi\rho\nu WR \left[1 + \frac{3}{16}Re \right] \tag{5.20}$$

It is readily shown that Equation (5.19) reduces to (5.11) as $Re \to 0$. The corresponding solution for the Hadamard-Rybczynski boundary conditions is not known to the author; its validity would be more questionable since, unlike the case of Stokes' boundary conditions, the inertial terms $u_j\partial u_i/\partial x_j$ are not identically zero on the surface of the bubble.

More recently Proudman and Pearson (1957) and Kaplun and Lagerstrom (1957) showed that Oseen's solution is, in fact, the first term obtained when the method of matched asymptotic expansions is used in an attempt to patch together consistent asymptotic solutions of the full Navier-Stokes equations for both the near field close to the sphere and the far field. They also obtained the next term in the expression for the drag force.

$$F_1 = -6\pi\rho\nu WR \left[1 + \frac{3}{16}Re + \frac{9}{160}Re^2 ln\left(\frac{Re}{2} \right) + 0(Re^2) \right] \tag{5.21}$$

The additional term leads to an error of 1% at $Re = 0.3$ and does not, therefore, have much practical consequence.

The most notable feature of the Oseen solution is that the geometry of the streamlines depends on the Reynolds number. The downstream flow is *not* a mirror image of the upstream flow as in the Stokes or potential flow solutions. Indeed, closer examination of the Oseen solution reveals that, downstream of the sphere, the streamlines are further apart and the flow is slower than in the equivalent upstream location. Furthermore, this effect increases with Reynolds number. These features of the Oseen solution are entirely consistent with experimental observations and represent the initial development of a wake behind the body.

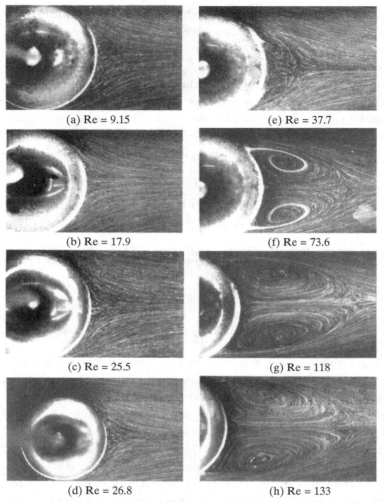

(a) Re = 9.15   (e) Re = 37.7

(b) Re = 17.9   (f) Re = 73.6

(c) Re = 25.5   (g) Re = 118

(d) Re = 26.8   (h) Re = 133

Figure 5.4 Streamlines of steady flow (from left to right) past a sphere at various Reynolds numbers (from Taneda 1956, reproduced by permission of the author).

The flow past a sphere at Reynolds numbers between about 0.5 and several thousand has proven intractable to analytical methods though numerical solutions are numerous. Experimentally, it is found that a recirculating zone (or vortex ring) develops close to the rear stagnation point at about $Re = 30$ (see Taneda 1956 and Figure 5.4). With further increase in the Reynolds number this recirculating zone or wake expands. Defining locations on the surface by the angle from the front stagnation point, the separation point moves forward from about 130° at $Re = 100$ to about 115° at $Re = 300$. In the process the wake reaches a diameter comparable to that of the sphere when $Re \approx 130$. At this point the flow becomes unstable and the ring vortex that makes up the wake begins to oscillate (Taneda 1956). However, it continues to be attached to the sphere until about $Re = 500$ (Torobin and Gauvin 1959).

At Reynolds numbers above about 500, vortices begin to be shed and then convected downstream. The frequency of vortex shedding has not been studied as extensively as in the case of a circular cylinder and seems to vary more with Reynolds number. In terms of the conventional Strouhal number, St, defined as

$$St = 2fR/W \tag{5.22}$$

the vortex shedding frequencies, f, that Moller (1938) observed correspond to a range of St varying from 0.3 at $Re = 1000$ to about 1.8 at $Re = 5000$. Furthermore, as Re increases above 500 the flow develops a fairly steady "near-wake" behind which vortex shedding forms an unsteady and increasingly turbulent "far-wake." This process continues until, at a value of Re of the order of 1000, the flow around the sphere and in the near-wake again becomes quite steady. A recognizable boundary layer has developed on the front of the sphere and separation settles down to a position about $84°$ from the front stagnation point. Transition to turbulence occurs on the free shear layer, which defines the boundary of the near-wake and moves progressively forward as the Reynolds number increases. The flow is similar to that of the top picture in Figure 5.2. Then the events described in the previous section occur with further increase in the Reynolds number.

Since the Reynolds number range between 0.5 and several hundred can often pertain in multiphase flows, one must resort to an empirical formula for the drag force in this regime. A number of empirical results are available; for example, Klyachko (1934) recommends

$$F_1 = -6\pi\rho\nu WR \left[1 + \frac{Re^{\frac{2}{3}}}{6} \right] \tag{5.23}$$

which fits the data fairly well up to $Re \approx 1000$. At $Re = 1$ the factor in the square brackets is 1.167, whereas the same factor in Equation (5.20) is 1.187. On the other hand, at $Re = 1000$, the two factors are respectively 17.7 and 188.5.

5.4 Marangoni Effects

As a postscript to the steady, viscous flows of the last section, it is of interest to introduce and describe the forces that a bubble may experience due to gradients in the surface tension, S, over the surface. These are called Marangoni effects. The gradients in the surface tension can be caused by a number of different factors. For example, gradients in the temperature, solvent concentration, or electric potential can create gradients in the surface tension. The "thermocapillary" effects due to temperature gradients have been explored by a number of investigators (for example, Young, Goldstein, and Block 1959) because of their importance in several technological contexts. For most of the range of temperatures, the surface tension decreases linearly with temperature, reaching zero at the critical point. Consequently, the controlling thermophysical property, dS/dT, is readily identified and more or less constant for any given fluid. Some typical data for dS/dT is presented in

Table 5.1. *Values of the temperature gradient of the surface*
tension, $-dS/dT$, for pure liquid/vapor interfaces (in
$kg/s^2 K$).

Water	2.02×10^{-4}	Methane	1.84×10^{-4}
Hydrogen	1.59×10^{-4}	Butane	1.06×10^{-4}
Helium-4	1.02×10^{-4}	Carbon Dioxide	1.84×10^{-4}
Nitrogen	1.92×10^{-4}	Ammonia	1.85×10^{-4}
Oxygen	1.92×10^{-4}	Toluene	0.93×10^{-4}
Sodium	0.90×10^{-4}	Freon-12	1.18×10^{-4}
Mercury	3.85×10^{-4}	Uranium Dioxide	1.11×10^{-4}

Table 5.1 and reveals a remarkably uniform value for this quantity for a wide range of liquids.

Surface tension gradients affect free surface flows because a gradient, dS/ds, in a direction, s, tangential to a surface clearly requires that a shear stress act in the negative s direction in order that the surface be in equilibrium. Such a shear stress would then modify the boundary conditions (for example, the Hadamard-Rybczynski conditions used in the preceding section), thus altering the flow and the forces acting on the bubble.

As an example of the Marangoni effect, we will examine the steady motion of a spherical bubble in a viscous fluid when there exists a gradient of the temperature (or other controlling physical property), dT/dx_1, in the direction of motion (see Figure 5.1). We must first determine whether the temperature (or other controlling property) is affected by the flow. It is illustrative to consider two special cases from a spectrum of possibilities. The first and simplest special case, which is not so relevant to the thermocapillary phenomenon, is to assume that $T = (dT/dx_1)x_1$ throughout the flow field so that, on the surface of the bubble,

$$\left(\frac{1}{R} \frac{dS}{d\theta} \right)_{r=R} = -\sin\theta \left(\frac{dS}{dT} \right) \left(\frac{dT}{dx_1} \right) \tag{5.24}$$

Much more realistic is the assumption that thermal conduction dominates the heat transfer ($\nabla^2 T = 0$) and that there is no heat transfer through the surface of the bubble. Then it follows from the solution of Laplace's equation for the conductive heat transfer problem that

$$\left(\frac{1}{R} \frac{dS}{d\theta} \right)_{r=R} = -\frac{3}{2} \sin\theta \left(\frac{dS}{dT} \right) \left(\frac{dT}{dx_1} \right) \tag{5.25}$$

The latter is the solution presented by Young, Goldstein, and Block (1959), but it differs from Equation (5.24) only in terms of the effective value of dS/dT. Here we shall employ Equation (5.25) since we focus on thermocapillarity, but other possibilities such as Equation (5.24) should be borne in mind.

For simplicity we will continue to assume that the bubble remains spherical. This assumption implies that the surface tension differences are small compared with the absolute level of S and that the stresses normal to the surface are entirely dominated by the surface tension.

With these assumptions the tangential stress boundary condition for the spherical bubble becomes

$$\rho_L \nu_L \left(\frac{\partial u_\theta}{\partial r} - \frac{u_\theta}{r} \right)_{r=R} + \frac{1}{R} \left(\frac{dS}{d\theta} \right)_{r=R} = 0 \qquad (5.26)$$

and this should replace the Hadamard-Rybczynski condition of zero shear stress that was used in the preceding section. Applying Equation (5.26) with Equation (5.25) and the usual kinematic condition, $(u_r)_{r=R} = 0$, to the general solution of the preceding section leads to

$$A = -\frac{R^4}{4\rho_L \nu_L} \frac{dS}{dx_1}; \quad B = \frac{WR}{2} + \frac{R^2}{4\rho_L \nu_L} \frac{dS}{dx_1} \qquad (5.27)$$

and consequently, from Equation (5.14), the force acting on the bubble becomes

$$F_1 = -4\pi \rho_L \nu_L WR - 2\pi R^2 \frac{dS}{dx_1} \qquad (5.28)$$

In addition to the normal Hadamard-Rybczynski drag (first term), we can identify a Marangoni force, $2\pi R^2 (dS/dx_1)$, acting on the bubble in the direction of *decreasing* surface tension. Thus, for example, the presence of a uniform temperature gradient, dT/dx_1, would lead to an additional force on the bubble of magnitude $2\pi R^2 (-dS/dT)(dT/dx_1)$ in the direction of the warmer fluid since the surface tension decreases with temperature. Such thermocapillary effects have been observed and measured by Young, Goldstein, and Block (1959) and others.

Finally, we should comment on a related effect caused by surface contaminants that increase the surface tension. When a bubble is moving through liquid under the action, say, of gravity, convection may cause contaminants to accumulate on the downstream side of the bubble. This will create a positive $dS/d\theta$ gradient which, in turn, will generate an effective shear stress acting in a direction opposite to the flow. Consequently, the contaminants tend to immobilize the surface. This will cause the flow and the drag to change from the Hadamard-Rybczynski solution to the Stokes solution for zero tangential velocity. The effect is more pronounced for smaller bubbles since, for a given surface tension difference, the Marangoni force becomes larger relative to the buoyancy force as the bubble size decreases. Experimentally, this means that surface contamination usually results in Stokes drag for spherical bubbles smaller than a certain size and in Hadamard-Rybczynski drag for spherical bubbles larger than that size. Such a transition is observed in experiments measuring the rise velocity of bubbles as, for example, in the Haberman and Morton (1953) experiments discussed in more detail in Section 5.12. The effect has been analyzed in the more complex hydrodynamic case at higher Reynolds numbers by Harper, Moore, and Pearson (1967).

5.5 Molecular Effects

Though only rarely important in the context of bubbles, there are some effects that can be caused by the molecular motions in the surrounding fluid. We briefly list some of these here.

When the mean free path of the molecules in the surrounding fluid, λ, becomes comparable with the size of the particles, the flow will clearly deviate from the continuum models, which are only relevant when $\lambda \ll R$. The Knudsen number, $Kn = \lambda/2R$, is used to characterize these circumstances, and Cunningham (1910) showed that the first-order correction for small but finite Knudsen number leads to an additional factor, $(1 + 2AKn)$, in the Stokes drag for a spherical particle. The numerical factor, A, is roughly a constant of order unity (see, for example, Green and Lane 1964).

When the impulse generated by the collision of a single fluid molecule with the particle is large enough to cause significant change in the particle velocity, the resulting random motions of the particle are called "Brownian motion" (Einstein 1956). This leads to diffusion of solid particles suspended in a fluid. Einstein showed that the diffusivity, D, of this process is given by

$$D = kT/6\pi\mu R \tag{5.29}$$

where k is Boltzmann's constant. It follows that the typical *rms* displacement, λ, of the particle in a time, t, is given by

$$\lambda = (kTt/3\pi\mu R)^{\frac{1}{2}} \tag{5.30}$$

Brownian motion is usually only significant for micron- and sub-micron-sized particles. The example quoted by Einstein is that of a 1 μm diameter particle in water at $17°C$ for which the typical displacement during one second is $0.8\ \mu m$.

A third, related phenomenon is the reponse of a particle to the collisions of molecules when there is a significant temperature gradient in the fluid. Then the impulses imparted to the particle by molecular collisions on the hot side of the particle will be larger than the impulses on the cold side. The particle will therefore experience a net force driving it in the direction of the colder fluid. This phenomenon is known as *thermophoresis* (see, for example, Davies 1966). A similar phenomenon known as *photophoresis* occurs when a particle is subjected to nonuniform radiation. One could, of course, include in this list the Bjerknes forces described in Section 4.10 since they constitute *sonophoresis*.

5.6 Unsteady Particle Motions

Having reviewed the steady motion of a particle relative to a fluid, we must now consider the consequences of unsteady relative motion in which either the particle or the fluid or both are accelerating. The complexities of fluid acceleration are delayed until the next section. First we shall consider the simpler circumstance in which the fluid is either at rest or has a steady uniform streaming motion (U = constant) far from the particle. Clearly the second case is readily reduced to the first by a simple Galilean transformation and it will be assumed that this has been accomplished.

In the ideal case of unsteady inviscid potential flow, it can then be shown by using the concept of the total kinetic energy of the fluid that the force on a rigid particle in an incompressible flow is given by F_i, where

$$F_i = -M_{ij} \frac{dV_j}{dt} \tag{5.31}$$

where M_{ij} is called the *added mass matrix* (or tensor) though the name "induced inertia tensor" used by Batchelor (1967) is, perhaps, more descriptive. The reader is referred to Sarpkaya and Isaacson (1981), Yih (1969), or Batchelor (1967) for detailed descriptions of such analyses. The above mentioned methods also show that M_{ij} for any finite particle can be obtained from knowledge of several *steady* potential flows. In fact,

$$M_{ij} = \frac{\rho}{2} \int_{\substack{volume \\ of\ fluid}} u_{ik} u_{jk}\, d(volume) \tag{5.32}$$

where the integration is performed over the entire volume of the fluid. The velocity field, u_{ij}, is the fluid velocity in the i direction caused by the *steady* translation of the particle with unit velocity in the j direction. Note that this means that M_{ij} is necessarily a symmetric matrix. Furthermore, it is clear that particles with planes of symmetry will not experience a force perpendicular to that plane when the direction of acceleration is parallel to that plane. Hence if there is a plane of symmetry perpendicular to the k direction, then for $i \neq k$, $M_{ki} = M_{ik} = 0$, and the only off-diagonal matrix elements that can be nonzero are M_{ij}, $j \neq k$, $i \neq k$. In the special case of the sphere *all* the off-diagonal terms will be zero.

Tables of some available values of the diagonal components of M_{ij} are given by Sarpkaya and Isaacson (1981) who also summarize the experimental results, particularly for planar flows past cylinders. Other compilations of added mass results can be found in Kennard (1967), Patton (1965), and Brennen (1982). Some typical values for three-dimensional particles are listed in Table 5.2. The uniform diagonal value for a sphere (often referred to simply as the added mass of a sphere) is $2\rho\pi R^3/3$ or one-half the displaced mass of fluid. This value can readily be obtained from Equation (5.32) using the steady flow results given in Equations (5.7) to (5.10). In general, of course, there is *no* special relation between the added mass and the displaced mass. Consider, for example, the case of the infinitely thin plate or disc with zero displaced mass which has a finite added mass in the direction normal to the surface. Finally, it should be noted that the literature contains little, if any, information on off-diagonal components of added mass matrices.

Now consider the application of these potential flow results to real viscous flows at high Reynolds numbers (the case of low Reynolds number flows will be discussed in Section 5.8). Significant doubts about the applicability of the added masses calculated from potential flow analysis would be justified because of the experience of D'Alembert's paradox for steady potential flows and the substantial difference

Table 5.2. *Added masses (diagonal terms in M_{ij}) for some three-dimensional bodies (particles): (T) Potential flow calculations, (E) Experimental data from Patton (1965).*

Particle		Matrix Element	Value		
Sphere (T)		M_{ii}	$\frac{2}{3}\rho C\pi R^3$		
Disc (T)		M_{11}	$\frac{8}{3}\rho C R^3$		
Ellipsoids (T)		$M_{ii} = K_{ii}\frac{4}{3}\rho C\pi ab^2$	a/b	K_{11}	$K_{22}(K_{33})$
			2	0.209	0.702
			5	0.059	0.895
			10	0.021	0.960
Sphere near Plane Wall (T) ($R/H \ll 1$)		$M_{ii} = K_{ii}\frac{4}{3}\rho C\pi R^3$	$K_{11} = \frac{1}{2}(1 + \frac{3}{8}\frac{R^3}{H^3} + \ldots)$ $K_{22} = \frac{1}{2}(1 + \frac{3}{16}\frac{R^3}{H^3} + \ldots)$ $K_{33} = K_{22}$		
Sphere near Free Surface (E) ($R/H \ll 1$)		$M_{ii} = K_{ii}\frac{4}{3}\rho C\pi R^3$	H/R	K_{11}	
			8.0	0.52	
			4.0	0.59	
			2.0	0.54	
			1.0	0.44	
			0.0	0.25	

between the streamlines of the potential and actual flows. Furthermore, analyses of experimental results will require the separation of the "added mass" forces from the viscous drag forces. Usually this is accomplished by heuristic summation of the two forces so that

$$F_i = -M_{ij}\frac{dV_j}{dt} - \frac{1}{2}\rho A C_{ij}|V_j|V_j \tag{5.33}$$

where C_{ij} is a lift and drag coefficient matrix and A is a typical cross-sectional area for the body. This is known as Morison's equation (see Morison et al. 1950).

Actual unsteady high Reynolds number flows are more complicated and not necessarily compatible with such simple superposition. This is reflected in the fact that the coefficients, M_{ij} and C_{ij}, appear from the experimental results to be not only functions of Re but also functions of the reduced time or frequency of the unsteady motion. Typically experiments involve either oscillation of a body in a fluid or acceleration from rest. The most extensively studied case involves planar flow past a cylinder (for example, Keulegan and Carpenter 1958), and a detailed review of this data is included in Sarkaya and Isaacson (1981). For oscillatory motion of the cylinder with velocity amplitude, U_M, and period, t^*, the coefficients are functions of both the Reynolds number, $Re = 2U_M R/\nu$, and the reduced period or Keulegan-Carpenter number, $Kc = U_M t^*/2R$. When the amplitude, $U_M t^*$, is less than about $10R$ ($Kc < 5$), the inertial effects dominate and M_{ii} is only a little less than its potential flow value over a wide range of Reynolds numbers ($10^4 < Re < 10^6$). However, for larger values of Kc, M_{ii} can be substantially smaller than this and, in some range of Re and Kc, may actually be negative. The values of C_{ii} (the drag coefficient) that are deduced from experiments are also a complicated function of Re and Kc. The behavior of the coefficients is particularly pathological when the reduced period, Kc, is close to that of vortex shedding (Kc of the order of 10). Large transverse or "lift" forces can be generated under these circumstances. To the author's knowledge, detailed investigations of this kind have not been made for a spherical body, but one might expect the same qualitative phenomena to occur.

5.7 Unsteady Potential Flow

In general, a particle moving in any flow other than a steady uniform stream will experience fluid accelerations, and it is therefore necessary to consider the structure of the equation governing the particle motion under these circumstances. Of course, this will include the special case of acceleration of a particle in a fluid at rest (or with a steady streaming motion). As in the earlier sections we shall confine the detailed solutions to those for a spherical particle or bubble. Furthermore, we consider only those circumstances in which both the particle and fluid acceleration are in one direction, chosen for convenience to be the x_1 direction. The effect of an external force field such as gravity will be omitted; it can readily be inserted into any of the solutions that follow by the addition of the conventional buoyancy force.

All the solutions discussed are obtained in an accelerating frame of reference *fixed* in the center of the fluid particle. Therefore, if the velocity of the particle in some original, noninertial coordinate system, x_i^*, was $V(t)$ in the x_1^* direction, the Navier-Stokes equations in the new frame, x_i, fixed in the particle center are

$$\frac{\partial u_i}{\partial t} + u_j \frac{\partial u_i}{\partial x_j} = -\frac{1}{\rho}\frac{\partial P}{\partial x_i} + \nu \frac{\partial^2 u_i}{\partial x_j \partial x_j} \tag{5.34}$$

where the pseudo-pressure, P, is related to the actual pressure, p, by

$$P = p + \rho x_1 \frac{dV}{dt} \tag{5.35}$$

Here the conventional time derivative of $V(t)$ is denoted by d/dt, but it should be noted that in the original x_i^* frame it implies a Lagrangian derivative following the particle. As before, the fluid is assumed incompressible (so that continuity requires $\partial u_i/\partial x_i = 0$) and Newtonian. The velocity that the fluid would have at the x_i origin in the absence of the particle is then $W(t)$ in the x_1 direction. It is also convenient to define the quantities r, θ, u_r, u_θ as shown in Figure 5.1 and the Stokes streamfunction as in Equations (5.6). In some cases we shall also be able to consider the unsteady effects due to growth of the bubble so the radius is denoted by $R(t)$.

First consider inviscid potential flow for which Equations (5.34) may be integrated to obtain the Bernoulli equation

$$\frac{\partial \phi}{\partial t} + \frac{P}{\rho} + \frac{1}{2}(u_\theta^2 + u_r^2) = constant \tag{5.36}$$

where ϕ is a velocity potential ($u_i = \partial\phi/\partial x_i$) and ψ must satisfy the equation

$$L\psi = 0 \quad \text{where} \quad L \equiv \frac{\partial^2}{\partial r^2} + \frac{\sin\theta}{r^2}\frac{\partial}{\partial\theta}\left(\frac{1}{\sin\theta}\frac{\partial}{\partial\theta}\right) \tag{5.37}$$

This is of course the same equation as in steady flow and has harmonic solutions, only five of which are necessary for present purposes:

$$\psi = \sin^2\theta\left\{-\frac{Wr^2}{2} + \frac{D}{r}\right\} + \cos\theta\sin^2\theta\left\{\frac{2Ar^3}{3} - \frac{B}{r^2}\right\} + E\cos\theta \tag{5.38}$$

$$\phi = \cos\theta\left\{-Wr + \frac{D}{r^2}\right\} + \left(\cos^2\theta - \frac{1}{3}\right)\left\{Ar^2 + \frac{B}{r^3}\right\} + \frac{E}{r} \tag{5.39}$$

$$u_r = \cos\theta\left\{-W - \frac{2D}{r^3}\right\} + \left(\cos^2\theta - \frac{1}{3}\right)\left\{2Ar - \frac{3B}{r^4}\right\} - \frac{E}{r^2} \tag{5.40}$$

$$u_\theta = -\sin\theta\left\{-W + \frac{D}{r^3}\right\} - 2\cos\theta\sin\theta\left\{Ar + \frac{B}{r^4}\right\} \tag{5.41}$$

The first part, which involves W and D, is identical to that for steady translation. The second, involving A and B, will provide the fluid velocity gradient in the x_1 direction, and the third, involving E, permits a time-dependent particle (bubble) radius. The W and A terms represent the fluid flow in the absence of the particle, and the D, B, and E terms allow the boundary condition

$$(u_r)_{r=R} = \frac{dR}{dt} \tag{5.42}$$

to be satisfied provided

$$D = -\frac{WR^3}{2}, \quad B = \frac{2AR^5}{3}, \quad E = -R^2\frac{dR}{dt} \tag{5.43}$$

In the absence of the particle the velocity of the fluid at the origin, $r = 0$, is simply $-W$ in the x_1 direction and the gradient of the velocity $\partial u_1/\partial x_1 = 4A/3$. Hence A is determined from the fluid velocity gradient in the original frame as

$$A = \frac{3}{4}\frac{\partial U}{\partial x_1^*} \tag{5.44}$$

Now the force, F_1, on the bubble in the x_1 direction is given by

$$F_1 = -2\pi R^2 \int_0^\pi p \sin\theta \cos\theta d\theta \qquad (5.45)$$

which upon using Equations (5.35), (5.36), and (5.39) to (5.41) can be integrated to yield

$$\frac{F_1}{2\pi R^2 \rho} = -\frac{D}{Dt}(WR) - \frac{4}{3}RWA + \frac{2}{3}R\frac{dV}{dt} \qquad (5.46)$$

Reverting to the original coordinate system and using τ as the sphere volume for convenience ($\tau = 4\pi R^3/3$), one obtains

$$F_1 = -\frac{1}{2}\rho\tau\frac{dV}{dt^*} + \frac{3}{2}\rho\tau\frac{DU}{Dt^*} + \frac{1}{2}\rho(U-V)\frac{d\tau}{dt^*} \qquad (5.47)$$

where the two Lagrangian time derivatives are defined by

$$\frac{D}{Dt^*} \equiv \frac{\partial}{\partial t^*} + U\frac{\partial}{\partial x_1^*} \qquad (5.48)$$

$$\frac{d}{dt^*} \equiv \frac{\partial}{\partial t^*} + V\frac{\partial}{\partial x_1^*} \qquad (5.49)$$

Equation 5.47 is an important result, and care must be taken not to confuse the different time derivatives contained in it. Note that in the absence of bubble growth, of viscous drag, and of body forces, the equation of motion that results from setting $F_1 = m_p dV/dt^*$ is

$$\left(1 + \frac{2m_p}{\rho\tau}\right)\frac{dV}{dt^*} = 3\frac{DU}{Dt^*} \qquad (5.50)$$

where m_p is the mass of the "particle." Thus for a massless bubble the acceleration of the bubble is three times the fluid acceleration.

In a more comprehensive study of unsteady potential flows Symington (1978) has shown that the result for more general (i.e., noncolinear) accelerations of the fluid and particle is merely the vector equivalent of Equation (5.47):

$$F_i = -\frac{1}{2}\rho\tau\frac{dV_i}{dt^*} + \frac{3}{2}\rho\tau\frac{DU_i}{Dt^*} + \frac{1}{2}\rho(U_i - V_i)\frac{d\tau}{dt^*} \qquad (5.51)$$

where

$$\frac{d}{dt^*} = \frac{\partial}{\partial t^*} + V_j\frac{\partial}{\partial x_j^*}; \quad \frac{D}{Dt^*} = \frac{\partial}{\partial t^*} + U_j\frac{\partial}{\partial x_j^*} \qquad (5.52)$$

The first term in Equation (5.51) represents the conventional added mass effect due to the particle acceleration. The factor 3/2 in the second term due to the fluid acceleration may initially seem surprising. However, it is made up of two components:

1. $\frac{1}{2}\rho dV_i/dt^*$, which is the added mass effect of the fluid acceleration
2. $\rho\tau DU_i/Dt^*$, which is a "buoyancy"-like force due to the pressure gradient associated with the fluid acceleration.

The last term in Equation (5.51) is caused by particle (bubble) volumetric growth, $d\tau/dt^*$, and is similar in form to the force on a source in a uniform stream.

Now it is necessary to ask how this force given by Equation (5.51) should be used in the practical construction of an equation of motion for a particle. Frequently, a viscous drag force F_i^D, is quite arbitrarily added to F_i to obtain some total "effective" force on the particle. Drag forces, F_i^D, with the conventional forms

$$F_i^D = C_D \cdot \frac{1}{2}\rho|U_i - V_i|(U_i - V_i)\pi R^2 \qquad (Re \gg 1) \qquad \cdot (5.53)$$

$$F_i^D = 6\pi\mu(U_i - V_i)R \qquad (Re \ll 1) \qquad (5.54)$$

have both been employed in the literature. It is, however, important to recognize that there is no fundamental analytical justification for such superposition of these forces. At high Reynolds numbers, we noted in the last section that experimentally observed added masses are indeed quite close to those predicted by potential flow within certain parametric regimes, and hence the superposition has some experimental justification. At low Reynolds numbers, it is improper to use the results of the potential flow analysis. The appropriate analysis under these circumstances is examined in the next section.

5.8 Unsteady Stokes Flow

In order to elucidate some of the issues raised in the last section, it is instructive to examine solutions for the unsteady flow past a sphere in low Reynolds number Stokes flow. In the asymptotic case of zero Reynolds number, the solution of Section 5.3 is unchanged by unsteadiness, and hence the solution at any instant in time is identical to the steady-flow solution for the same particle velocity. In other words, since the fluid has no inertia, it is always in static equilibrium. Thus the instantaneous force is identical to that for the steady flow with the same $V_i(t)$.

The next step is therefore to investigate the effects of small but nonzero inertial contributions. The Oseen solution provides some indication of the effect of the *convective* inertial terms, $u_j\partial u_i/\partial x_j$, in steady flow. Here we investigate the effects of the *unsteady* inertial term, $\partial u_i/\partial t$. Ideally it would be best to include *both* the $\partial u_i/\partial t$ term *and* the Oseen approximation to the convective term, $U\partial u_i/\partial x$. However, the resulting *unsteady* Oseen flow is sufficiently difficult that only small-time expansions for the impulsively started motions of droplets and bubbles exist in the literature (Pearcey and Hill 1956).

Consider, therefore the *unsteady* Stokes equations in the absence of the convective inertial terms:

$$\rho\frac{\partial u_i}{\partial t} = -\frac{\partial P}{\partial x_i} + \mu\frac{\partial^2 u_i}{\partial x_j\partial x_j} \qquad (5.55)$$

Since both the equations and the boundary conditions used below are linear in u_i, we need only consider colinear particle and fluid velocities in one direction, say x_1. The solution to the general case of noncolinear particle and fluid velocities and accelerations may then be obtained by superposition. As in Section 5.7 the colinear problem is solved by first transforming to an accelerating coordinate frame, x_i, fixed in the center of the particle so that $P = p + \rho x_1 dV/dt$. Elimination of P by taking the curl of Equation (5.55) leads to

$$\left(L - \frac{1}{\nu}\frac{\partial}{\partial t} \right) L\psi = 0 \tag{5.56}$$

where L is the same operator as defined in Equation (5.37). Guided by both the steady Stokes flow and the unsteady potential flow solution, one can anticipate a solution of the form

$$\psi = \sin^2\theta\, f(r,t) + \cos\theta\sin^2\theta\, g(r,t) + \cos\theta\, h(t) \tag{5.57}$$

plus other spherical harmonic functions. The first term has the form of the steady Stokes flow solution; the last term would be required if the particle were a growing spherical bubble. After substituting Equation (5.57) into Equation (5.56), the equations for f, g, h are

$$\left(L_1 - \frac{1}{\nu}\frac{\partial}{\partial t} \right) L_1 f = 0 \qquad \text{where} \qquad L_1 \equiv \frac{\partial^2}{\partial r^2} - \frac{2}{r^2} \tag{5.58}$$

$$\left(L_2 - \frac{1}{\nu}\frac{\partial}{\partial t} \right) L_2 g = 0 \qquad \text{where} \qquad L_2 \equiv \frac{\partial^2}{\partial r^2} - \frac{6}{r^2} \tag{5.59}$$

$$\left(L_0 - \frac{1}{\nu}\frac{\partial}{\partial t} \right) L_0 h = 0 \qquad \text{where} \qquad L_0 \equiv \frac{\partial^2}{\partial r^2} \tag{5.60}$$

Moreover, the form of the expression for the force, F_1, on the spherical particle (or bubble) obtained by evaluating the stresses on the surface and integrating is

$$\frac{F_1}{\frac{4}{3}\rho\pi R^3} = \frac{dV}{dt} + \left[\frac{1}{r}\frac{\partial^2 f}{\partial r\partial t} + \frac{\nu}{r}\left\{ \frac{2}{r^2}\frac{\partial f}{\partial r} + \frac{2}{r}\frac{\partial^2 f}{\partial r^2} - \frac{\partial^3 f}{\partial r^3} \right\} \right]_{r=R} \tag{5.61}$$

It transpires that this is *independent* of g or h. Hence only the solution to Equation (5.58) for $f(r,t)$ need be sought in order to find the force on a spherical particle, and the other spherical harmonics that might have been included in Equation (5.57) are now seen to be unnecessary.

Fourier or Laplace transform methods may be used to solve Equation (5.58) for $f(r,t)$, and we choose Laplace transforms. The Laplace transforms for the relative velocity $W(t)$, and the function $f(r,t)$ are denoted by $\hat{W}(s)$ and $\hat{f}(r,s)$:

$$\hat{W}(s) = \int_0^\infty e^{-st} W(t)\,dt; \qquad \hat{f}(r,s) = \int_0^\infty e^{-st} f(r,t)\,dt \tag{5.62}$$

Then Equation (5.58) becomes

$$(L_1 - \alpha^2)L_1\hat{f} = 0 \tag{5.63}$$

where $\alpha^2 = s/\nu$, and the solution after application of the condition that $\hat{u}_1(s,t)$ far from the particle be equal to $\hat{W}(s)$ is

$$\hat{f} = -\frac{\hat{W}r^2}{2} + \frac{A(s)}{r} + B(s)\left(\frac{1}{r} + \alpha\right)e^{-\alpha r} \tag{5.64}$$

where $\alpha = (s/\nu)^{\frac{1}{2}}$ and A and B are as yet undetermined functions of s. Their determination requires application of the boundary conditions on $r = R$. In terms of A and B the Laplace transform of the force $\hat{F}_1(s)$ is

$$\frac{\hat{F}_1}{\frac{4}{3}\pi R^3 \rho} = \frac{d\hat{V}}{dt} + \left[\frac{s}{r}\frac{\partial\hat{f}}{\partial r} + \frac{\nu}{R}\left\{-\frac{4\hat{W}}{r} + \frac{8A}{r^4} + CBe^{-\alpha r}\right\}\right]_{r=R} \tag{5.65}$$

where

$$C = \alpha^4 + \frac{3\alpha^3}{r} + \frac{3\alpha^2}{r^2} + \frac{8\alpha}{r^3} + \frac{8}{r^4} \tag{5.66}$$

The classical solution (see Landau and Lifshitz 1959) is for a solid sphere (i.e., constant R) using the no-slip (Stokes) boundary condition for which

$$f(R,t) = \left.\frac{\partial f}{\partial r}\right|_{r=R} = 0 \tag{5.67}$$

and hence

$$A = +\frac{\hat{W}R^3}{2} + \frac{3\hat{W}R\nu}{2s}\{1 + \alpha R\}; \quad B = -\frac{3\hat{W}R\nu}{2s}e^{\alpha R} \tag{5.68}$$

so that

$$\frac{\hat{F}_1}{\frac{4}{3}\pi R^3 \rho} = \frac{d\hat{V}}{dt} - \frac{3}{2}s\hat{W} - \frac{9\nu\hat{W}}{2R^2} - \frac{9\nu^{\frac{1}{2}}}{2R}s^{\frac{1}{2}}\hat{W} \tag{5.69}$$

For a motion starting at rest at $t = 0$ the inverse Laplace transform of this yields

$$\frac{F_1}{\frac{4}{3}\pi R^3 \rho} = \frac{dV}{dt} - \frac{3}{2}\frac{dW}{dt} - \frac{9\nu}{2R^2}W - \frac{9}{2R}\left(\frac{\nu}{\pi}\right)^{\frac{1}{2}}\int_0^t \frac{dW(\tilde{t})}{d\tilde{t}}\frac{d\tilde{t}}{(t-\tilde{t})^{\frac{1}{2}}} \tag{5.70}$$

where $\tilde{t}$ is a dummy time variable. This result must then be written in the original coordinate framework with $W = V - U$ and can be generalized to the noncolinear case by superposition so that

$$F_i = -\frac{1}{2}\tau\rho\frac{dV_i}{dt^*} + \frac{3}{2}\tau\rho\frac{dU_i}{dt^*} + \frac{9\tau\mu}{2R^2}(U_i - V_i) + \frac{9\tau\rho}{2R}\left(\frac{\nu}{\pi}\right)^{\frac{1}{2}}$$

$$\times \int_0^{t^*} \frac{d(U_i - V_i)}{d\tilde{t}}\frac{d\tilde{t}}{(t^* - \tilde{t})^{\frac{1}{2}}} \tag{5.71}$$

where d/dt^* is the Lagrangian time derivative following the particle. This is then the general force on the particle or bubble in unsteady Stokes flow when the Stokes boundary conditions are applied.

Compare this result with that obtained from the potential flow analysis, Equation (5.51) with τ taken as constant. It is striking to observe that the coefficients of the added mass terms involving dV_i/dt^* and dU_i/dt^* are identical to those of the potential flow solution. On superficial examination it might be noted that dU_i/dt^* appears in Equation (5.71) whereas DU_i/Dt^* appears in (5.51); the difference is, however, of order $W_j \partial U_i/dx_j$ and terms of this order have already been dropped from the equation of motion on the basis that they were negligible compared with the temporal derivatives like $\partial W_i/\partial t$. Hence it is inconsistent with the initial assumption to distinguish between d/dt^* and D/Dt^* in the present unsteady Stokes flow solution.

The term $9\nu W/2R^2$ in Equation (5.71) is, of course, the steady Stokes drag. The new phenomenon introduced by this analysis is contained in the last term of Equation (5.71). This is a fading memory term that is often named the Basset term after one of its identifiers (Basset 1888). It results from the fact that additional vorticity created at the solid particle surface due to relative acceleration diffuses into the flow and creates a temporary perturbation in the flow field. Like all diffusive effects it produces an $\omega^{\frac{1}{2}}$ term in the equation for oscillatory motion.

Before we conclude this section, comment should be included on two other analytical results. Morrison and Stewart (1976) have considered the case of a spherical bubble for which the Hadamard-Rybczynski boundary conditions rather than the Stokes conditions are applied. Then, instead of the conditions of Equation (5.67), the conditions for zero normal velocity and zero shear stress on the surface require that

$$f(R,t) = \left[\frac{\partial^2 f}{\partial r^2} - \frac{2}{r}\frac{\partial f}{\partial r}\right]_{r=R} = 0 \tag{5.72}$$

and hence in this case (see Morrison and Stewart 1976)

$$A(s) = +\frac{\hat{W}R^3}{2} + \frac{3\hat{W}R(1+\alpha R)}{\alpha^2(3+\alpha R)} \; ; \; B(s) = -\frac{3\hat{W}Re^{+\alpha R}}{\alpha^2(3+\alpha R)} \tag{5.73}$$

so that

$$\frac{\hat{F}_1}{\frac{4}{3}\pi R^3 \rho} = \frac{d\hat{V}}{dt} - \frac{9\hat{W}\nu}{R^2} - \frac{3}{2}\hat{W}s + \frac{6\nu\hat{W}}{R^2\left\{1+s^{\frac{1}{2}}R/3\nu^{\frac{1}{2}}\right\}} \tag{5.74}$$

The inverse Laplace transform of this for motion starting at rest at $t=0$ is

$$\frac{F_1}{\frac{4}{3}\pi R^3 \rho} = \frac{dV}{dt} - \frac{3}{2}\frac{dW}{dt} - \frac{3\nu W}{R^2} \tag{5.75}$$

$$-\frac{6\nu}{R^2}\int_0^t \frac{dW(\tilde{t})}{d\tilde{t}}\exp\left\{\frac{9\nu(t-\tilde{t})}{R^2}\right\}\mathrm{erfc}\left\{\left(\frac{9\nu(t-\tilde{t})}{R^2}\right)^{\frac{1}{2}}\right\}d\tilde{t}$$

Comparing this with the solution for the Stokes conditions, we note that the first two terms are unchanged and the third term is the expected Hadamard-Rybczynski steady drag term (see Equation (5.16)). The last term is significantly different from the Basset term in Equation (5.71) but still represents a receding memory.

The second interesting case is that for unsteady Oseen flow, which essentially consists of attempting to solve the Navier-Stokes equations with the convective initial terms approximated by $U_j \partial u_i / \partial x_j$. Pearcey and Hill (1956) have examined the small-time behavior of droplets and bubbles started from rest when this term is included in the equations.

5.9 Growing or Collapsing Bubbles

We now return to the discussion of higher Re flow and specifically address the effects due to bubble growth or collapse. A bubble that grows or collapses close to a boundary may undergo translation due to the asymmetry induced by that boundary. A relatively simple example of the analysis of this class of flows is the case of the growth or collapse of a spherical bubble near a plane boundary, a problem first solved by Herring (1941) (see also Davies and Taylor 1942, 1943). Assuming that the only translational motion of the bubble (with velocity, W) is perpendicular to the plane boundary, the geometry of the bubble and its image in the boundary will be as shown in Figure 5.5. For convenience, we define additional polar coooordinates, $(\breve{r}, \breve{\theta})$, with origin at the center of the image bubble. Assuming inviscid, irrotational flow, Herring (1941) and Davies and Taylor (1943) constructed the velocity potential, ϕ, near the bubble by considering an expansion in terms of R/h where h is the distance of the bubble center from the boundary. Neglecting all terms that are of order R^3/h^3 or higher, the velocity potential can be obtained by superposing the individual contributions from the bubble source/sink, the image source/sink, the bubble translation dipole, the image dipole, and one correction factor described below. This combination yields

$$\phi = -\frac{R^2 \dot{R}}{r} - \frac{WR^3 \cos\theta}{2r^2} \pm \left[-\frac{R^2 \dot{R}}{\breve{r}} + \frac{WR^3 \cos\breve{\theta}}{2\breve{r}^2} - \frac{R^5 \dot{R}\cos\theta}{8h^2 r^2} \right] \tag{5.76}$$

The first and third terms are the source/sink contributions from the bubble and the image respectively. The second and fourth terms are the dipole contributions due

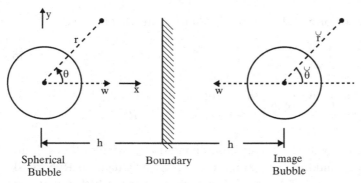

Figure 5.5 Schematic of a bubble undergoing growth or collapse close to a plane boundary. The associated translational velocity is denoted by W.

to the translation of the bubble and the image. The last term arises because the source/sink in the bubble needs to be displaced from the bubble center by an amount $R^3/8h^2$ normal to the wall in order to satisfy the boundary condition on the surface of the bubble to order R^2/h^2. All other terms of order R^3/h^3 or higher are neglected in this analysis assuming that the bubble is sufficiently far from the boundary so that $h \gg R$. Finally, the sign choice on the last three terms of Equation (5.76) is as follows: the upper, positive sign pertains to the case of a solid boundary and the lower, negative sign provides an approximate solution for a free surface boundary.

It remains to use this solution to determine the translational motion, $W(t)$, normal to the boundary. This is accomplished by invoking the condition that there is no net force on the bubble. Using the unsteady Bernoulli equation and the velocity potential and fluid velocities obtained from Equation (5.76), Davies and Taylor (1943) evaluate the pressure at the bubble surface and thereby obtain an expression for the force, F_x, on the bubble in the x direction:

$$F_x = -\frac{2\pi}{3} \left\{ \frac{d}{dt}\left(R^3 W\right) \pm \frac{3}{4}\frac{R^2}{h^2}\frac{d}{dt}\left(R^3 \frac{dR}{dt}\right) \right\} \qquad (5.77)$$

Adding the effect of buoyancy due to a component, g_x, of the gravitational acceleration in the x direction, Davies and Taylor then set the total force equal to zero and obtain the following equation of motion for $W(t)$:

$$\frac{d}{dt}\left(R^3 W\right) \pm \frac{3}{4}\frac{R^2}{h^2}\frac{d}{dt}\left(R^3 \frac{dR}{dt}\right) + \frac{4\pi R^3 g_x}{3} = 0 \qquad (5.78)$$

In the absence of gravity this corresponds to the equation of motion first obtained by Herring (1941). Many of the studies of growing and collapsing bubbles near boundaries have been carried out in the context of underwater explosions (see Cole 1948). An example illustrating the solution of Equation (5.78) and the comparison with experimental data is included in Figure 5.6 taken from Davies and Taylor (1943).

Another application of this analysis is to the translation of cavitation bubbles near walls. Here the motivation is to understand the development of impulsive loads on the solid surface (see Section 3.6), and therefore the primary focus is on bubbles close to the wall so that the solution described above is of limited value since it requires $h \gg R$. However, as discussed in Section 3.5, considerable progress has been made in recent years in developing analytical methods for the solution of the inviscid free surface flows of bubbles near boundaries. One of the concepts that is particularly useful in determining the direction of bubble translation is based on a property of the flow first introduced by Kelvin (see Lamb 1932) and called the Kelvin impulse. This vector property applies to the flow generated by a finite particle or bubble in a fluid; it is denoted by I_{Ki} and defined by

$$I_{Ki} = \rho \int_{S_B} \phi n_i dS \qquad (5.79)$$

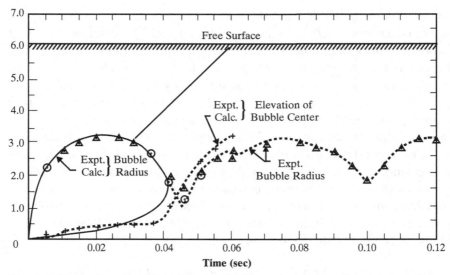

Figure 5.6 Data from Davies and Taylor (1943) on the mean radius and central elevation of a bubble in oil generated by a spark-initiated explosion of 1.32×10^6 *ergs* situated $6.05\ cm$ below the free surface. The two measures of the bubble radius are one half of the horizontal span ($\triangle$) and one quarter of the sum of the horizontal and vertical spans ($\odot$). Theoretical calculations using Equation (5.78) are indicated by the solid lines.

where ϕ is the velocity potential of the irrotational flow, S_B is the surface of the bubble, and n_i is the outward normal at that surface (defined as positive into the bubble). If one visualizes a bubble in a fluid at rest, then the Kelvin impulse is the impulse that would have to be applied to the bubble in order to generate the motions of the fluid related to the bubble motion. Benjamin and Ellis (1966) were the first to demonstrate the value of this property in determining the interaction between a growing or collapsing bubble and a nearby boundary (see also Blake and Gibson 1987).

5.10 Equation of Motion

In a multiphase flow with a very dilute discrete phase the fluid forces discussed in Sections 5.1 to 5.8 will determine the motion of the particles that constitute that discrete phase. In this section we discuss the implications of some of the fluid force terms. The equation that determines the particle velocity, V_i, is generated by equating the total force, F_i^T, on the particle to $m_p dV_i/dt^*$. Consider the motion of a spherical particle or (bubble) of mass m_p and volume τ (radius R) in a *uniformly* accelerating fluid. The simplest example of this is the vertical motion of a particle under gravity, g, in a pool of otherwise quiescent fluid. Thus the results will be written in terms of the buoyancy force. However, the same results apply to motion generated by any uniform acceleration of the fluid, and hence g can be interpreted as a general uniform fluid acceleration (dU/dt). This will also allow some tentative conclusions to be drawn concerning the relative motion of a particle in the nonuniformly accelerating fluid situations that can occur in general multiphase flow. For the motion of a sphere at small relative Reynolds number, $Re_W \ll 1$ (where $Re_W = 2WR/\nu$ and W is the typical magnitude of the relative velocity), only the forces due to buoyancy and the weight of the particle need be added to F_i as given by Equations (5.71) or (5.75) in order to obtain F_i^T. This addition is simply given by $(\rho\tau - m_p)g_i$ where g is a vector in the vertically upward direction with magnitude equal to the acceleration due to gravity. On the other hand, at high relative Reynolds numbers, $Re_W \gg 1$, one must resort to a more heuristic approach in which the fluid forces given by Equation (5.51) are supplemented by drag (and lift) forces given by $\frac{1}{2}\rho A C_{ij}|W_j|W_j$ as in Equation (5.33). In either case it is useful to nondimensionalize the resulting equation of motion so that the pertinent nondimensional parameters can be identified.

Examine first the case in which the relative velocity, W (defined as positive in the direction of the acceleration, g, and therefore positive in the vertically upward direction of the rising bubble or sedimenting particle), is sufficiently small so that the relative Reynolds number is much less than unity. Then, using the Stokes boundary conditions, the equation governing W may be obtained from Equation (5.70) as

$$w + \frac{dw}{dt_*} + \left\{\frac{9}{\pi(1 + 2m_p/\rho\tau)}\right\}^{\frac{1}{2}} \int_0^{t_*} \frac{dw}{d\tilde{t}}\frac{d\tilde{t}}{(t_* - \tilde{t})^{\frac{1}{2}}} = 1 \qquad (5.80)$$

where the dimensionless time

$$t_* = t/t_R \quad \text{and} \quad t_R = R^2(1 + 2m_p/\rho\tau)/9\nu \qquad (5.81)$$

and $w = W/W_\infty$ where W_∞ is the steady terminal velocity given by

$$W_\infty = 2R^2g(1 - m_p/\rho\tau)/9\nu \qquad (5.82)$$

In the absence of the Basset term the solution of Equation (5.80) is simply

$$w = 1 - e^{-t/t_R} \qquad (5.83)$$

and the typical response time, t_R, is called the relaxation time for particle velocity (see, for example, Rudinger 1969). In the general case that includes the Basset term the dimensionless solution, $w(t_*)$, of Equation (5.80) depends only on the parameter $m_p/\rho\tau$ (particle mass/displaced fluid mass) appearing in the Basset term. Indeed, the dimensionless Equation (5.80) clearly illustrates the fact that the Basset term is much less important for solid particles in a gas where $m_p/\rho\tau \gg 1$ than it is for bubbles in a liquid where $m_p/\rho\tau \ll 1$. Note also that for initial conditions of zero relative velocity ($w(0) = 0$) the small-time solution of Equation (5.80) takes the form

$$w = t_* - \frac{2}{\pi^{\frac{1}{2}}\{1 + 2m_p/\rho\tau\}^{\frac{1}{2}}}t_*^{\frac{3}{2}} + \dots \qquad (5.84)$$

Hence the initial acceleration at $t = 0$ is given dimensionally by $2g(1 - m_p/\rho\tau)/(1 + 2m_p/\rho\tau)$ or $2g$ in the case of a massless bubble and $-g$ in the case of a heavy

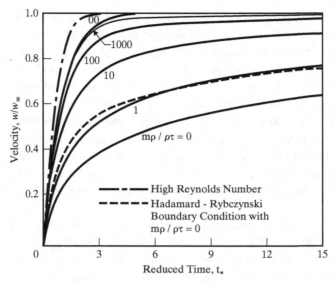

Figure 5.7 The velocity, W, of a particle released from rest at $t_* = 0$ in a quiescent fluid and its approach to terminal velocity, W_∞. Horizontal axis is a dimensionless time defined in text. Solid lines represent the low Reynolds number solutions for various particle mass/displaced mass ratios, $m_p/\rho\tau$, and the Stokes boundary condition. The dashed line is for the Hadamard-Rybczynski boundary condition and $m_p/\rho\tau = 0$. The dash-dot line is the high Reynolds number result; note that t_* is nondimensionalized differently in that case.

solid particle in a gas where $m_p \gg \rho\tau$. Note also that the effect of the Basset term is to *reduce* the acceleration of the relative motion, thus increasing the time required to achieve terminal velocity.

Numerical solutions of the form of $w(t_*)$ for various $m_p/\rho\tau$ are shown in Figure 5.7 where the delay caused by the Basset term can be clearly seen. In fact in the later stages of approach to the terminal velocity the Basset term dominates over the added mass term, (dw/dt_*). The integral in the Basset term becomes approximately $2t_*^{\frac{1}{2}} dw/dt_*$ so that the final approach to $w = 1$ can be approximated by

$$w = 1 - C\exp\left\{ -t_*^{\frac{1}{2}} \Big/ \left(\frac{9}{\pi\{1+2m_p/\rho\tau\}} \right)^{\frac{1}{2}} \right\} \tag{5.85}$$

where C is a constant. As can be seen in Figure 5.7, the result is a much slower approach to W_∞ for small $m_p/\rho\tau$ than for larger values of this quantity.

The case of a bubble with Hadamard-Rybczynski boundary conditions is very similar except that

$$W_\infty = R^2 g(1 - m_p/\rho\tau)/3\nu \tag{5.86}$$

and the equation for $w(t_*)$ is

$$w + \frac{3}{2}\frac{dw}{dt_*} + 2\int_0^{t_*} \frac{dw}{d\tilde{t}}\Gamma(t_* - \tilde{t})d\tilde{t} = 1 \tag{5.87}$$

where the function, $\Gamma(\xi)$, is given by

$$\Gamma(\xi) = \exp\left\{\left(1 + \frac{2m_p}{\rho\tau}\right)\xi\right\} \mathrm{erfc}\left\{\left(\left(1 + \frac{2m_p}{\rho\tau}\right)\xi\right)^{\frac{1}{2}}\right\} \qquad (5.88)$$

For the purposes of comparison the form of $w(t_*)$ for the Hadamard-Rybczynski boundary condition with $m_p/\rho\tau = 0$ is also shown in Figure 5.7. Though the altered Basset term leads to a more rapid approach to terminal velocity than occurs for the Stokes boundary condition, the difference is not qualitatively significant.

If the terminal Reynolds number is much greater than unity then, in the absence of particle growth, Equation (5.51) heuristically supplemented with a drag force of the form of Equation (5.53) leads to the following equation of motion for unidirectional motion:

$$w^2 + \frac{dw}{dt_*} = 1 \qquad (5.89)$$

where $w = W/W_\infty, t_* = t/t_R$,

$$t_R = (1 + 2m_p/\rho\tau)(2R/3C_Dg(1 - m_p/\tau\rho))^{\frac{1}{2}} \qquad (5.90)$$

and

$$W_\infty = [8Rg(1 - m_p/\rho\tau)/3C_D]^{\frac{1}{2}} \qquad (5.91)$$

The solution to Equation (5.89) for $w(0) = 0$,

$$w = \tanh t_* \qquad (5.92)$$

is also shown in Figure 5.7 though, of course, t_* has a different definition in this case.

For the purposes of reference in Section 5.12 note that, if we define a Reynolds number, Re, Froude number, Fr, and drag coefficient, C_D, by

$$Re = \frac{2W_\infty R}{\nu}; \quad Fr = \frac{W_\infty}{[2Rg(1 - m_p/\rho\tau)]^{\frac{1}{2}}} \qquad (5.93)$$

then the expressions for the terminal velocities, W_∞, given by Equations (5.82), (5.86), and (5.91) can be written as

$$Fr = (Re/18)^{\frac{1}{2}}, \ Fr = (Re/12)^{\frac{1}{2}}, \quad \text{and} \quad Fr = (4/3C_D)^{\frac{1}{2}} \qquad (5.94)$$

respectively. Indeed, dimensional analysis of the governing Navier-Stokes equations requires that the general expression for the terminal velocity can be written as

$$F(Re, Fr) = 0 \qquad (5.95)$$

or, alternatively, if C_D is defined as $4/3Fr^2$, then it could be written as

$$F^*(Re, C_D) = 0 \qquad (5.96)$$

5.11 Magnitude of Relative Motion

Qualitative estimates of the magnitude of the relative motion in multiphase flows can be made from the analyses of the last section. Consider a general steady fluid flow characterized by a velocity, U, and a typical dimension, ℓ; it may, for example, be useful to visualize the flow in a converging nozzle of length, ℓ, and mean axial velocity, U. A particle in this flow will experience a typical fluid acceleration (or effective g) of U^2/ℓ for a typical time given by ℓ/U and hence will develop a velocity, W, relative to the fluid. In many practical flows it is necessary to determine the maximum value of W (denoted by W_M) that could develop under these circumstances. To do so, one must first consider whether the available time, ℓ/U, is large or small compared with the typical time, t_R, required for the particle to reach its terminal velocity as given by Equation (5.81) or (5.90). If $t_R \ll \ell/U$ then W_M is given by Equation (5.82), (5.86), or (5.91) for W_∞ and qualitative estimates for W_M/U would be

$$\left(1 - \frac{m_p}{\rho\tau}\right)\left(\frac{UR}{\nu}\right)\left(\frac{R}{\ell}\right) \quad \text{and} \quad \left(1 - \frac{m_p}{\rho\tau}\right)^{\frac{1}{2}} \frac{1}{C_D^{\frac{1}{2}}}\left(\frac{R}{\ell}\right)^{\frac{1}{2}} \tag{5.97}$$

when $WR/\nu \ll 1$ and $WR/\nu \gg 1$ respectively. We refer to this as the quasistatic regime. On the other hand, if $t_T \gg \ell/U$, W_M can be estimated as $W_\infty \ell/U t_R$ so that W_M/U is of the order of

$$\frac{2(1 - m_p/\rho\tau)}{(1 + 2m_p/\rho\tau)} \tag{5.98}$$

for all WR/ν. This is termed the transient regime.

In practice, WR/ν will not be known in advance. The most meaningful quantities that can be evaluated prior to any analysis are a Reynolds number, UR/ν, based on flow velocity and particle size, a size parameter

$$X = \frac{R}{\ell}\left|1 - \frac{m_p}{\rho\tau}\right| \tag{5.99}$$

and the parameter

$$Y = \left|1 - \frac{m_p}{\rho\tau}\right| \Big/ \left(1 + \frac{2m_p}{\rho\tau}\right) \tag{5.100}$$

The resulting regimes of relative motion are displayed graphically in Figure 5.8. The transient regime in the upper right-hand sector of the graph is characterized by large relative motion, as suggested by Equation (5.98). The quasistatic regimes for $WR/\nu \gg 1$ and $WR/\nu \ll 1$ are in the lower right- and left-hand sectors respectively. The shaded boundaries between these regimes are, of course, approximate and are functions of the parameter Y, which must have a value in the range $0 < Y < 1$. As one proceeds deeper into either of the quasistatic regimes, the magnitude of the relative velocity, W_M/U, becomes smaller and smaller. Thus, homogeneous flows (see Chapter 6) in which the relative motion is neglected require that *either* $X \ll Y^2$ *or* $X \ll Y/(UR/\nu)$. Conversely, if either of these conditions is violated, relative motion must be included in the analysis.

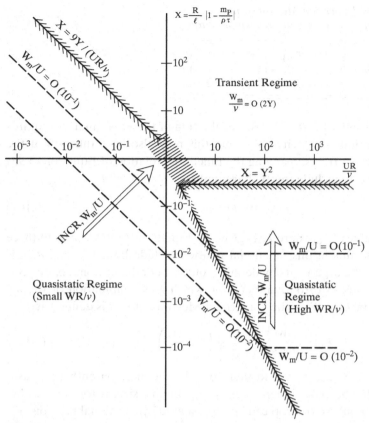

Figure 5.8 Schematic of the various regimes of relative motion between a particle and the surrounding flow.

5.12 Deformation Due to Translation

In the case of bubbles, drops, or deformable particles it has thus far been tacitly assumed that their shape is known and constant. Since the fluid stresses due to translation may deform such a particle, we must now consider not only the parameters governing the deformation but also the consequences in terms of the translation velocity and the shape. We concentrate here on bubbles and drops in which surface tension, S, acts as the force restraining deformation. However, the reader will realize that there would exist a similar analysis for deformable elastic particles. Furthermore, the discussion will be limited to the case of *steady* translation, caused by gravity, g. Clearly the results could be extended to cover translation due to fluid acceleration by using an effective value of g as indicated in the last section.

The characteristic force maintaining the sphericity of the bubble or drop is given by SR. Deformation will occur when the characteristic anisotropy in the fluid forces approaches SR; the magnitude of the anisotropic fluid force will be given by $\mu W_\infty R$ for $W_\infty R/\nu \ll 1$ or by $\rho W_\infty^2 R^2$ for $W_\infty R/\nu \gg 1$. Thus defining a Weber number, $We = 2\rho W_\infty^2 R/S$, deformation will occur when We/Re approaches unity for $Re \ll 1$ or

Table 5.3. *Values of the Haberman-Morton numbers,*
$Hm = g\mu^4/\rho S^3$, *for various liquids at normal temperatures.*

Filtered Water	0.25×10^{-10}	Turpentine	2.41×10^{-9}
Methyl Alcohol	0.89×10^{-10}	Olive Oil	7.16×10^{-3}
Mineral Oil	1.45×10^{-2}	Syrup	0.92×10^{6}

when *We* approaches unity for $Re \gg 1$. But evaluation of these parameters requires knowledge of the terminal velocity, W_∞, and this may also be a function of the shape. Thus one must start by expanding the functional relation of Equation (5.95) which determines W_∞ to include the Weber number:

$$F(Re, We, Fr) = 0 \tag{5.101}$$

This relation determines W_∞ where *Fr* is given by Equations (5.93). Since all three dimensionless coefficients in this functional relation include both W_∞ and *R*, it is simpler to rearrange the arguments by defining another nondimensional parameter known as the Haberman-Morton number, *Hm*, which is a combination of *We*, *Re*, and *Fr* but does not involve W_∞. The Haberman-Morton number is defined as

$$Hm = \frac{We^3}{Fr^2 Re^4} = \frac{g\mu^4}{\rho S^3}\left(1 - \frac{m_p}{\rho\tau}\right) \tag{5.102}$$

In the case of a bubble, $m_p \ll \rho\tau$ and therefore the factor in parenthesis is usually omitted. Then *Hm* becomes independent of the bubble size. It follows that the terminal velocity of a bubble or drop can be represented by functional relation

$$F(Re, Hm, Fr) = 0 \quad \text{or} \quad F^*(Re, Hm, C_D) = 0 \tag{5.103}$$

and we shall confine the following discussion to the nature of this relation for bubbles ($m_p \ll \rho\tau$).

Some values for the Haberman-Morton number (with $m_p/\rho\tau = 0$) for various saturated liquids are shown in Figure 5.9; other values are listed in Table 5.3. Note that for all but the most viscous liquids, *Hm* is much less than unity. It is, of course, possible to have fluid accelerations much larger than *g*; however, this is unlikely to cause *Hm* values greater than unity in practical multiphase flows of most liquids.

Having introduced the Haberman-Morton number, we can now identify the conditions for departure from sphericity. For low Reynolds numbers ($Re \ll 1$) the terminal velocity will be given by $Re \propto Fr^2$. Then the shape will deviate from spherical when $We \geq Re$ or, using $Re \propto Fr^2$ and $Hm = We^3 Fr^{-2} Re^{-4}$, when

$$Re \geq Hm^{-\frac{1}{2}} \tag{5.104}$$

Thus if $Hm < 1$ all bubbles for which $Re \ll 1$ will remain spherical. However, there are some unusual circumstances in which $Hm > 1$ and then there will be a range of *Re*, namely $Hm^{-\frac{1}{2}} < Re < 1$, in which significant departure from sphericity might occur.

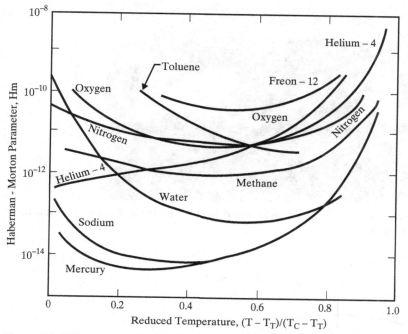

Figure 5.9 Values of the Haberman-Morton parameter, *Hm*, for various pure substances as a function of reduced temperature.

For high Reynolds numbers ($Re \gg 1$) the terminal velocity is given by $Fr \approx O(1)$ and distortion will occur if $We > 1$. Using $Fr = 1$ and $Hm = We^3 Fr^{-2} Re^{-4}$ it follows that departure from sphericity will occur when

$$Re \gg Hm^{-\frac{1}{4}} \qquad (5.105)$$

Consequently, in the common circumstances in which $Hm < 1$, there exists a range of Reynolds numbers, $Re < Hm^{-\frac{1}{4}}$, in which sphericity is maintained; nonspherical shapes occur when $Re > Hm^{-\frac{1}{4}}$. For $Hm > 1$ departure from sphericity has already occurred at $Re < 1$ as discussed above.

Experimentally, it is observed that the initial departure from sphericity causes ellipsoidal bubbles that may oscillate in shape and have oscillatory trajectories (Hartunian and Sears 1957). As the bubble size is further increased to the point at which $We \approx 20$, the bubble acquires a new asymptotic shape, known as a "spherical-cap bubble." A photograph of a typical spherical-cap bubble is shown in Figure 5.10; the notation used to describe the approximate geometry of these bubbles is sketched in Figure 5.11. Spherical-cap bubbles were first investigated by Davies and Taylor (1950), who observed that the terminal velocity is simply related to the radius of curvature of the cap, R_C, or to the equivalent volumetric radius, R_B, by

$$W_\infty = \frac{2}{3}(gR_C)^{\frac{1}{2}} = (gR_B)^{\frac{1}{2}} \qquad (5.106)$$

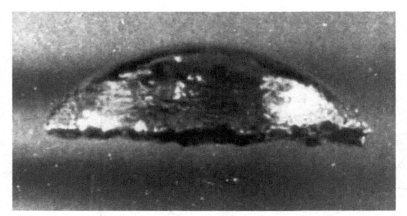

Figure 5.10 Photograph of a spherical cap bubble rising in water (from Davenport, Bradshaw, and Richardson 1967).

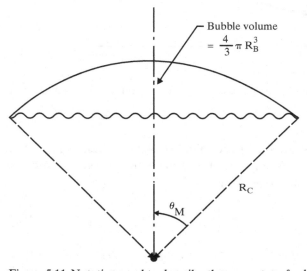

Figure 5.11 Notation used to describe the geometry of spherical cap bubbles.

Assuming a typical laminar drag coefficient of $C_D = 0.5$, a spherical solid particle with the same volume would have a terminal velocity,

$$W_\infty = (8gR_B/3C_D)^{\frac{1}{2}} = 2.3(gR_B)^{\frac{1}{2}} \tag{5.107}$$

which is substantially higher than the spherical-cap bubble. From Equation (5.106) it follows that the effective C_D for spherical-cap bubbles is 2.67 based on the area πR_B^2.

Wegener and Parlange (1973) have reviewed the literature on spherical-cap bubbles. Figure 5.12 is taken from from their review and shows that the value of $W_\infty/(gR_B)^{\frac{1}{2}}$ reaches a value of about 1 at a Reynolds number, $Re = 2W_\infty R_B/\nu$, of about 200 and, thereafter, remains fairly constant. Visualization of the flow reveals

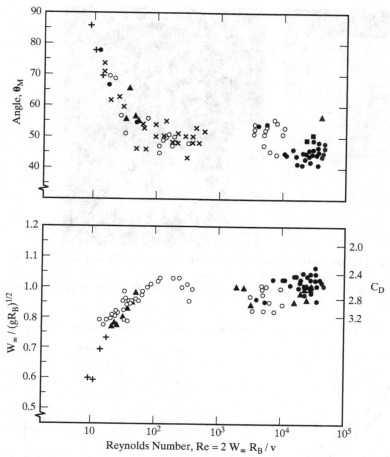

Figure 5.12 Data on the terminal velocity, $W_\infty/(gR_B)^{\frac{1}{2}}$, and the conical angle, θ_M, for spherical-cap bubbles studied by a number of different investigators (adapted from Wegener and Parlange 1973).

that, for Reynolds numbers less than about 360, the wake behind the bubble is laminar and takes the form of a toroidal vortex (similar to a Hill (1894) spherical vortex) shown in the left-hand photograph of Figure 5.13. The wake undergoes transition to turbulence about $Re = 360$, and bubbles at higher Re have turbulent wakes as illustrated in the right side of Figure 5.13. We should add that scuba divers have long observed that spherical-cap bubbles rising in the ocean seem to have a maximum size of the order of 30 *cm* in diameter. When they grow larger than this, they fission into two (or more) bubbles. However, the author has found no quantitative study of this fission process.

In closing, we note that the terminal velocities of the bubbles discussed here may be represented according to the functional relation of Equations (5.103) as a family of $C_D(Re)$ curves for various Hm. Figure 5.14 has been extracted from the experimental data of Haberman and Morton (1953) and shows the dependence of

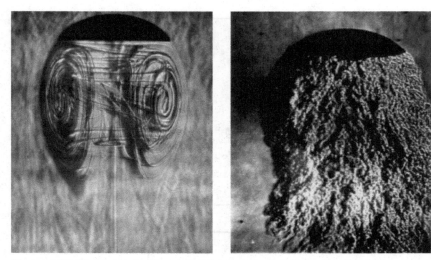

Figure 5.13 Flow visualizations of spherical-cap bubbles. On the left is a bubble with a laminar wake at $Re \approx 180$ (from Wegener and Parlange 1973) and, on the right, a bubble with a turbulent wake at $Re \approx 17,000$ (from Wegener, Sundell and Parlange 1971, reproduced with permission of the authors).

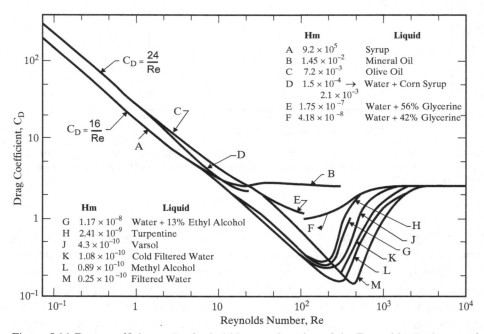

Figure 5.14 Drag coefficients, C_D, for bubbles as a function of the Reynolds number, Re, for a range of Haberman-Morton numbers, Hm, as shown. Data from Haberman and Morton (1953).

$C_D(Re)$ on Hm at intermediate Re. The curves cover the spectrum from the low Re spherical bubbles to the high Re spherical cap bubbles. The data demonstrate that, at higher values of Hm, the drag coefficient makes a relatively smooth transition from the low Reynolds number result to the spherical cap value of about 2.7. Lower values of *Hm* result in a deep minimum in the drag coefficient around a Reynolds number of about 200.

References

Basset, A.B. (1888). *A treatise on hydrodynamics*, **II**. Reprinted by Dover, NY, 1961.

Batchelor, G.K. (1967). *An introduction to fluid dynamics*. Cambridge Univ. Press.

Benjamin, T.B. and Ellis, A.T. (1966). The collapse of cavitation bubbles and the pressures thereby produced against solid boundaries. *Phil. Trans. Roy. Soc., London, Ser. A*, **260**, 221–240.

Blake, J.R. and Gibson, D.C. (1987). Cavitation bubbles near boundaries. *Ann. Rev. Fluid Mech.*, **19**, 99–124.

Brennen, C.E. (1982). A review of added mass and fluid inertial forces. *Naval Civil Eng. Lab., Port Hueneme, Calif., Report CR82.010*.

Cole, R.H. (1948). *Underwater explosions*. Princeton Univ. Press (reprinted by Dover, 1965).

Cunningham, E. (1910). On the velocity of steady fall of spherical particles through fluid medium. *Proc. Roy. Soc. A*, **83**, 357–365.

Davenport, W.G., Bradshaw, A.V., and Richardson, F.D. (1967). Behavior of spherical-cap bubbles in liquid metals. *J. Iron Steel Inst.*, **205**, 1034–1042.

Davies, C.N. (1966). *Aerosol science*. Academic Press, New York.

Davies, R.M. and Taylor, G.I. (1942). The vertical motion of a spherical bubble and the pressure surrounding it. In *The Scientific Papers of G.I.Taylor*, **III**, 320–336, edited by G.K.Batchelor, Cambridge Univ. Press.

Davies, R.M. and Taylor, G.I. (1943). The motion and shape of the hollow produced by an explosion in a liquid. In *The Scientific Papers of G.I.Taylor*, **III**, 337–353, edited by G.K.Batchelor, Cambridge Univ. Press.

Davies, R.M. and Taylor, G.I. (1950). The mechanics of large bubbles rising through extended liquids and through liquids in tubes. *Proc. Roy. Soc. A*, **200**, 375–390.

Einstein, A. (1956). *Investigations on the theory of Brownian movement*. Dover Publ., Inc., New York.

Green, H.L. and Lane, W.R. (1964). *Particulate clouds: dusts, smokes and mists*. E. and F.N. Spon Ltd., London.

Haberman, W.L. and Morton, R.K. (1953). An experimental investigation of the drag and shape of air bubbles rising in various liquids. *David Taylor Model Basin, Washington, Report No. 802*.

Hadamard, J. (1911). Movement permanent lent d'une sphere liquide et visqueuse dans un liquide visqueux. *Comptes Rendus*, **152**, 1735.

Harper, J.F., Moore, D.W. and Pearson, J.R.A. (1967). The effect of the variation of surface tension with temperature on the motion of bubbles and drops. *J. Fluid Mech.*, **27**, 361–366.

Hartunian, R.A. and Sears, W.R. (1957). On the instability of small gas bubbles moving uniformly in various liquids. *J. Fluid Mech.*, **3**, 27–47.

Herring, C. (1941). The theory of the pulsations of the gas bubbles produced by an underwater explosion. *US Nat. Defence Res. Comm. Report*.

Hill, M.J.M. (1894). On a spherical vortex. *Phil. Trans. Roy. Soc., London, Ser. A.*, **185**, 213–245.

Kaplun, S. and Lagerstrom, P.A. (1957). Asymptotic expansions of Navier-Stokes solutions for small Reynolds numbers. *J. Math. Mech.*, **6**, 585–593.

Kennard, E.M. (1967). Irrotational flow of frictionless fluid, mostly of invariable density. *David Taylor Model Basin, Washington, Report No. 2299*.

Keulegan, G.H. and Carpenter, L.H. (1958). Forces on cylinders and plates in an oscillating fluid. *U.S. Nat. Bur. Standards J. Res.*, **60**, No. 5, 423–440.

Klyachko, L.S. (1934). Heating and ventilation. *USSR Journal Otopl. i Ventil.*, No.4.

Lamb, H. (1932). *Hydrodynamics*. Cambridge Univ. Press.

Landau, L.E. and Lifshitz, E.M. (1959). *Fluid Mechanics*. Pergamon Press, NY.

Moller, W. (1938). Experimentelle Untersuchungen zur Hydrodynamik der Kugel. *Physik. Z.*, **39**, 57–80.

Morrison, F.A. and Stewart, M.B. (1976). Small bubble motion in an accelerating liquid. *ASME J. Appl. Mech.*, **43**, 399–403.

Morison, J.R., O'Brien, M.P., Johnson, J.W., and Schaaf, S.A. (1950). The forces exerted by surface waves on piles. *AIME Trans., Petroleum Branch*, **189**, 149–154.

Oseen, C.W. (1910). Über die Stokessche Formel und über die verwandte Aufgabe in der Hydrodynamik. *Arkiv Mat., Astron. och Fysik*, **6**, No. 29.

Patton, K.T. (1965). Tables of hydrodynamic mass factors for translational motion. *ASME Paper, 65-WA/UNT-2*.

Pearcey, T. and Hill, G.W. (1956). The accelerated motion of droplets and bubbles. *Australian J. of Phys.*, **9**, 19–30.

Proudman, I. and Pearson, J.R.A. (1957). Expansions at small Reynolds number for the flow past a sphere and a circular cylinder. *J. Fluid Mech.*, **2**, 237–262.

Rudinger, G. (1969). Relaxation in gas-particle flow. In *Nonequilibrium flows. Part 1*, (ed: P.P.Wegener), Marcel Dekker, New York and London.

Rybzynski, W. (1911). Über die fortschreitende Bewegung einer flüssigen Kugel in einem zähen Medium. *Bull. Acad. Sci. Cracovie*, **A**, 40.

Sarpkaya, T. and Isaacson, M. (1981). *Mechanics of wave forces on offshore structures*. Van Nostrand Reinhold Co., NY.

Stokes, G.G. (1851). On the effect of the internal friction of fluids on the motion of pendulums. *Trans. Camb. Phil. Soc.*, **9**, Part II, 8–106.

Symington, W.A. (1978). Analytical studies of steady and non-steady motion of a bubbly liquid. *Ph.D. Thesis, Calif. Inst. of Tech.*

Taneda, S. (1956). Studies on wake vortices (III). Experimental investigation of the wake behind a sphere at low Reynolds number. *Rep. Res. Inst. Appl. Mech., Kyushu Univ.*, **4**, 99–105.

Torobin, L.B. and Gauvin, W.H. (1959). Fundamental aspects of solids-gas flow. Part II. The sphere wake in steady laminar fluids. *Canadian J. Chem. Eng.*, **37**, 167–176.

Wegener, P.P., Sundell, R.E., and Parlange, J.-Y. (1971). Spherical-cap bubbles rising in liquids. *Z. Flugwissenschaften*, **19**, 347–352.

Wegener, P.P. and Parlange, J.-Y. (1973). Spherical-cap bubbles. *Ann. Rev. Fluid Mech.*, **5**, 79–100.

Yih, C.-S. (1969). *Fluid mechanics*. McGraw-Hill Book Co.

Young, N.O., Goldstein, J.S., and Block, M.J. (1959). The motion of bubbles in a vertical temperature gradient. *J. Fluid Mech.*, **6**, 350–356.

6 Homogeneous Bubbly Flows

6.1 Introduction

When the concentration of bubbles in a flow exceeds some small value the bubbles will begin to have a substantial effect on the fluid dynamics of the suspending liquid. Analyses of the dynamics of this multiphase mixture then become significantly more complicated and important new phenomena may be manifest. In this chapter we discuss some of the analyses and phenomena that may occur in bubbly multiphase flow.

In the larger context of practical multiphase (or multicomponent) flows one finds a wide range of homogeneities, from those consisting of one phase (or component) that is very finely dispersed within the other phase (or component) to those that consist of two separate streams of the two phases (or components). In between are topologies that are less readily defined. The two asymptotic states are conveniently referred to as *homogeneous* and *separated* flow. One of the consequences of the topology is the extent to which relative motion between the phases can occur. It is clear that two different streams can readily travel at different velocities, and indeed such relative motion is an implicit part of the study of separated flows. On the other hand, it is clear from the results of Section 5.11 that any two phases could, in theory, be sufficiently well mixed and the disperse particle size sufficiently small so as to eliminate any significant relative motion. Thus the asymptotic limit of truly homogeneous flow precludes relative motion. Indeed, the term homogeneous flow is sometimes used to denote a flow with negligible relative motion. Many bubbly flows come close to this limit and can, to a first approximation, be considered to be homogeneous. In the present chapter we shall consider some of the properties of homogeneous bubbly flows.

In the absence of relative motion the governing mass and momentum conservation equations reduce to a form similar to those for single-phase flow. The effective mixture density, ρ, is defined by

$$\rho = \sum_N \alpha_N \rho_N \tag{6.1}$$

where α_N is the volume fraction of each of the N components or phases whose individual densities are ρ_N. Then the continuity and momentum equations for the homogeneous mixture are

$$\frac{\partial \rho}{\partial t} + \frac{\partial}{\partial x_j}(\rho u_j) = 0 \qquad (6.2)$$

$$\frac{\partial}{\partial t}(\rho u_i) + \frac{\partial}{\partial x_j}(\rho u_i u_j) = -\frac{\partial p}{\partial x_i} + \rho g_i \qquad (6.3)$$

in the absence of viscous effects. As in single-phase flows the existence of a barotropic relation, $p = f(\rho)$, would complete the system of equations. In some multiphase flows it is possible to establish such a barotropic relation, and this allows one to anticipate (with, perhaps, some minor modification) that the entire spectrum of phenomena observed in single-phase gas dynamics can be expected in such a two-phase flow. In this chapter we shall not dwell on this established body of literature. Rather, we shall confine attention to the identification of a barotropic relation (if any) and focus on some flows in which there are major departures from the conventional gas dynamic behavior.

From a thermodynamic point of view the existence of a barotropic relation, $p = f(\rho)$, and its associated sonic speed,

$$c = \left(\frac{dp}{d\rho}\right)^{\frac{1}{2}} \qquad (6.4)$$

implies that some thermodynamic property is considered to be held constant. In single-phase gas dynamics this quantity is usually the entropy and occasionally the temperature. In multiphase flows the alternatives are neither simple nor obvious. In single-phase gas dynamics it is commonly assumed that the gas is in thermodynamic equilibrium at all times. In multiphase flows it is usually the case that the two phases are *not* in thermodynamic equilibrium with each other. These are some of the questions one must address in considering an appropriate homogeneous flow model for a multiphase flow. We begin in the next section by considering the sonic speed of a two-phase or two-component mixture.

6.2 Sonic Speed

Consider an infinitesmal volume of a mixture consisting of a disperse phase denoted by the subscript A and a continuous phase denoted by the subscript B. For convenience assume the initial volume to be unity. Denote the initial densities by ρ_A and ρ_B and the initial pressure in the *continuous* phase by p_B. Surface tension, S, can be included by denoting the radius of the disperse phase particles by R. Then the initial pressure in the disperse phase is $p_A = p_B + 2S/R$.

Now consider that the pressure, p_A, is changed to $p_A + \delta p_A$ where the difference δp_A is infinitesmal. Any dynamics associated with the resulting fluid motions will be ignored for the moment. It is assumed that a new equilibrium state is achieved and that, in the process, a mass, δm, is transferred from the continuous to the disperse phase. It follows that the new disperse and continuous phase masses are $\rho_A \alpha_A + \delta m$ and $\rho_B \alpha_B - \delta m$ respectively where, of course, $\alpha_B = 1 - \alpha_A$. Hence the new disperse and continuous phase volumes are respectively

$$(\rho_A \alpha_A + \delta m) / \left[\rho_A + \frac{\partial \rho_A}{\partial p_A} \Big|_{QA} \delta p_A \right] \tag{6.5}$$

and

$$(\rho_B \alpha_B - \delta m) / \left[\rho_B + \frac{\partial \rho_B}{\partial p_B} \Big|_{QB} \delta p_B \right] \tag{6.6}$$

where the thermodynamic constraints QA and QB are, as yet, unspecified. Adding these together and subtracting unity, one obtains the change in total volume, δV, and hence the sonic velocity, c, as

$$c^{-2} = -\rho \frac{\delta V}{\delta p_B} \Big|_{\delta p_B \to 0} \tag{6.7}$$

$$c^{-2} = \rho \left[\frac{\alpha_A}{\rho_A} \frac{\partial \rho_A}{\partial p_A} \Big|_{QA} \frac{\delta p_A}{\delta p_B} + \frac{\alpha_B}{\rho_B} \frac{\partial \rho_B}{\partial p_B} \Big|_{QB} - \frac{(\rho_B - \rho_A)}{\rho_A \rho_B} \frac{\delta m}{\delta p_B} \right] \tag{6.8}$$

where, as defined in Equation (6.1), $\rho = \rho_A \alpha_A + \rho_B \alpha_B$. If one assumes that no disperse particles are created or destroyed, then the ratio $\delta p_A / \delta p_B$ may be determined by evaluating the new disperse particle size $R + \delta R$ commensurate with the new disperse phase volume and using the relation $\delta p_A = \delta p_B - \frac{2S}{R^2} \delta R$:

$$\frac{\delta p_A}{\delta p_B} = \left[1 - \frac{2S}{3 \alpha_A \rho_A R} \frac{\delta m}{\delta p_B} \right] / \left[1 - \frac{2S}{3 \rho_A R} \frac{\partial \rho_A}{\partial p_A} \Big|_{QA} \right] \tag{6.9}$$

Substituting this into Equation (6.8) and using, for convenience, the notation

$$\frac{1}{c_A^2} = \frac{\partial \rho_A}{\partial p_A} \Big|_{QA}; \quad \frac{1}{c_B^2} = \frac{\partial \rho_B}{\partial p_B} \Big|_{QB} \tag{6.10}$$

the result can be written as

$$\frac{1}{\rho c^2} = \frac{\alpha_B}{\rho_B c_B^2} + \frac{\left[\frac{\alpha_A}{\rho_A c_A^2} - \frac{\delta m}{\delta p_B} \left\{ \frac{1}{\rho_A} - \frac{1}{\rho_B} + \frac{2S}{3 \rho_A \rho_B c_A^2 R} \right\} \right]}{\left[1 - \frac{2S}{3 \rho_A c_A^2 R} \right]} \tag{6.11}$$

This is incomplete in several respects. First, appropriate thermodynamic constraints QA and QB must be identified. Second, some additional constraint is necessary to establish the relation $\delta m / \delta p_B$. But before entering into a discussion of appropriate practical choices for these constraints (see Section 6.3) several simpler versions of Equation (6.11) should be discussed.

We first observe that in the absence of any exchange of mass between the components the result reduces to

$$\frac{1}{\rho c^2} = \frac{\alpha_B}{\rho_B c_B^2} + \frac{\frac{\alpha_A}{\rho_A c_A^2}}{\left\{ 1 - \frac{2S}{3 \rho_A c_A^2 R} \right\}} \tag{6.12}$$

In most practical cases one can neglect the surface tension effect since $S \ll \rho_A c_A^2 R$ and Equation (6.12) becomes

$$\frac{1}{c^2} = \{\rho_A \alpha_A + \rho_B \alpha_B\} \left[\frac{\alpha_B}{\rho_B c_B^2} + \frac{\alpha_A}{\rho_A c_A^2} \right] \quad (6.13)$$

In other words, the acoustic impedance $1/\rho c^2$ for the mixture is simply given by the average of the acoustic impedance of the components weighted according to their volume fractions.

Perhaps the most dramatic effects occur when one of the components is a gas (subscript G), which is much more compressible than the other component (a liquid or solid, subscript L). In the absence of surface tension ($p = p_G = p_L$), according to Equation (6.13), it matters not whether the gas is the continuous or the disperse phase. Denoting α_G by α for convenience and assuming the gas is perfect and behaves polytropically according to $\rho_G^k \propto p$, Equation (6.13) may be written as

$$\frac{1}{c^2} = [\rho_L(1-\alpha) + \rho_G \alpha] \left[\frac{\alpha}{kp} + \frac{(1-\alpha)}{\rho_L c_L^2} \right] \quad (6.14)$$

This is the familiar form for the sonic speed in a two-component gas/liquid or gas/solid flow. In many applications $p/\rho_L c_L^2 \ll 1$ and hence this expression may be further simplified to

$$\frac{1}{c^2} = \frac{\alpha}{kp} [\rho_L(1-\alpha) + \rho_G \alpha] \quad (6.15)$$

Note however, that this approximation will not hold for small values of the gas volume fraction α.

Equation (6.14) and its special properties were first identified by Minnaert (1933). It clearly exhibits one of the most remarkable features of the sonic velocity of gas/liquid or gas/solid mixtures. The sonic velocity of the mixture can be very much smaller than that of either of its constituents. This is illustrated in Figure 6.1 where the speed of sound, c, in an air/water bubbly mixture is plotted against the air volume fraction, α. Results are shown for both isothermal ($k = 1$) and adiabatic ($k = 1.4$) bubble behavior using Equation (6.14) or (6.15), the curves for these two equations being indistinguishable on the scale of the figure. Note that sonic velocities as low as 20 m/s occur.

Also shown in Figure 6.1 is experimental data of Karplus (1958) and Gouse and Brown (1964). We shall see later (Section 6.8) that the dynamics of the bubble volume change cause the sound speed to be a function of the frequency. Data for sound frequencies of $1.0\ kHz$ and $0.5\ kHz$ are shown, as well as data extrapolated to zero frequency. The last should be compared with the analytical results presented here since the analysis of this section neglects bubble dynamic effects. Note that the data corresponds to the isothermal theory, indicating that the heat transfer between the bubbles and the liquid is sufficient to maintain the air in the bubbles at roughly constant temperature.

Further discussion of the acoustic characteristics of dilute bubbly mixtures is delayed until Section 6.8.

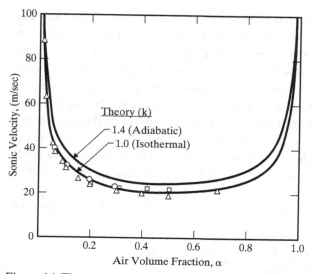

Figure 6.1 The sonic velocity in a bubbly air/water mixture at atmospheric pressure for $k = 1.0$ and 1.4. Experimental data presented is from Karplus (1958) and Gouse and Brown (1964) for frequencies of 1 kHz ($\odot$), 0.5 kHz ($\square$), and extrapolated to zero frequency ($\triangle$).

6.3 Sonic Speed with Change of Phase

Turning now to the behavior of a two-phase rather than two-component mixture, it is necessary not only to consider the additional thermodynamic constraint required to establish the mass exchange, δm, but also to reconsider the two thermodynamic constraints, QA and QB, which were implicit in the two-component analysis. These latter constraints were implicit in the choice of the polytropic index, k, for the gas and the choice of the sonic speed, c_L, for the liquid. Note that a nonisentropic choice for k (for example, $k = 1$) implies that heat is exchanged between the components, and yet this heat transfer process was not explicitly considered, nor was an overall thermodynamic contraint such as might be placed on the global change in entropy.

We shall see that the two-phase case requires more intimate knowledge of these factors because the results are more sensitive to the thermodynamic constraints. In an ideal, infinitely homogenized mixture of vapor and liquid the phases would everywhere be in such close proximity to each other that heat transfer between the phases would occur instantaneously. The entire mixture of vapor and liquid would then always be in thermodynamic equilibrium. Indeed, one model of the response of the mixture, called the *homogeneous equilibrium model*, assumes this to be the case. In practice, however, one seeks results for bubbly flows and mist flows in which heat transfer between the phases does not occur so readily. A second common model assumes zero heat transfer between the phases and is known as the *homogeneous frozen model*. In many circumstances the actual response lies somewhere between these extremes. A limited amount of heat transfer occurs between those portions of each phase that are close to the interface. In order to incorporate this in the analysis, we adopt an approach that includes the homogeneous equilibrium and homogeneous frozen responses as special cases but that requires a minor adjustment to the analysis of the last section in order to reflect the degree of thermal exchange between the phases. As in the last section the total mass of the phases A and B after application of the incremental pressure, δp, are $\rho_A \alpha_A + \delta m$ and $\rho_B \alpha_B - \delta m$, respectively. We now define the fractions of each phase, ϵ_A and ϵ_B which, because of their proximity to the interface, exchange heat and therefore approach thermodynamic equilibrium with each other. The other fractions $(1 - \epsilon_A)$ and $(1 - \epsilon_B)$ are assumed to be effectively insulated so that they behave isentropically. This is, of course, a crude simplification of the actual circumstances, but it permits qualitative assessment of practical flows.

It follows that the volumes of the four fractions following the incremental change in pressure, δp, are

$$\frac{(1 - \epsilon_A)(\rho_A \alpha_A + \delta m)}{\left[\rho_A + \delta p (\partial \rho_A / \partial p)_S\right]}; \quad \frac{\epsilon_A(\rho_A \alpha_A + \delta m)}{\left[\rho_A + \delta p (\partial \rho_A / \partial p)_E\right]}$$

$$\frac{(1 - \epsilon_B)(\rho_B \alpha_B - \delta m)}{\left[\rho_B + \delta p (\partial \rho_B / \partial p)_S\right]}; \quad \frac{\epsilon_B(\rho_B \alpha_B - \delta m)}{\left[\rho_B + \delta p (\partial \rho_B / \partial p)_E\right]} \tag{6.16}$$

where the subscripts S and E refer to isentropic and phase equilibrium derivatives, respectively. Then the change in total volume leads to the following modified form for Equation (6.11) in the absence of surface tension:

$$\frac{1}{\rho c^2} = (1 - \epsilon_A)\frac{\alpha_A}{\rho_A}\left(\frac{\partial \rho_A}{\partial p}\right)_S + \epsilon_A \frac{\alpha_A}{\rho_A}\left(\frac{\partial \rho_A}{\partial p}\right)_E + (1 - \epsilon_B)\frac{\alpha_B}{\rho_B}\left(\frac{\partial \rho_B}{\partial p}\right)_S$$

$$+ \epsilon_B \frac{\alpha_B}{\rho_B}\left(\frac{\partial \rho_B}{\partial p}\right)_E - \frac{\delta m}{\delta p}\left(\frac{1}{\rho_A} - \frac{1}{\rho_B}\right) \tag{6.17}$$

The exchange of mass, δm, is now determined by imposing the constraint that the entropy of the whole be unchanged by the perturbation. The entropy prior to δp is

$$\rho_A \alpha_A s_A + \rho_B \alpha_B s_B \tag{6.18}$$

where s_A and s_B are the specific entropies of the two phases. Following the application of δp, the entropy is

$$(1 - \epsilon_A)\{\rho_A \alpha_A + \delta m\} s_A + \epsilon_A \{\rho_A \alpha_A + \delta m\} \left\{s_A + \delta p (\partial s_A / \partial p)_E\right\}$$

$$+ (1 - \epsilon_B)\{\rho_B \alpha_B - \delta m\} s_B + \epsilon_B \{\rho_B \alpha_B - \delta m\} \left\{s_B + \delta p (\partial s_B / \partial p)_E\right\} \tag{6.19}$$

Equating (6.18) and (6.19) and writing the result in terms of the specific enthalpies h_A and h_B rather than s_A and s_B, one obtains

$$\frac{\delta m}{\delta p} = \frac{1}{(h_A - h_B)}\left[\epsilon_A \alpha_A \left\{1 - \rho_A \left(\frac{\partial h_A}{\partial p}\right)_E\right\} + \epsilon_B \alpha_B \left\{1 - \rho_B \left(\frac{\partial h_B}{\partial p}\right)_E\right\}\right] \tag{6.20}$$

Note that if the communicating fractions ϵ_A and ϵ_B were both zero, this would imply no exchange of mass. Thus $\epsilon_A = \epsilon_B = 0$ corresponds to the homogeneous frozen model (in which $\delta m = 0$) whereas $\epsilon_A = \epsilon_B = 1$ clearly yields the homogeneous equilibrium model.

Substituting Equation (6.20) into Equation (6.17) and rearranging the result, one can write

$$\frac{1}{\rho c^2} = \frac{\alpha_A}{p} \left[(1 - \epsilon_A)f_A + \epsilon_A g_A \right] + \frac{\alpha_B}{p} \left[(1 - \epsilon_B)f_B + \epsilon_B g_B \right] \tag{6.21}$$

where the quantities f_A, f_B, g_A, and g_B are purely thermodynamic properties of the two phases defined by

$$f_A = \left(\frac{\partial \ln \rho_A}{\partial \ln p} \right)_S ; \quad f_B = \left(\frac{\partial \ln \rho_B}{\partial \ln p} \right)_S \tag{6.22}$$

$$g_A = \left(\frac{\partial \ln \rho_A}{\partial \ln p} \right)_E + \left(\frac{1}{\rho_A} - \frac{1}{\rho_B} \right) \left(\rho_A h_A \frac{\partial \ln h_A}{\partial \ln p} - p \right)_E \bigg/ (h_A - h_B)$$

$$g_B = \left(\frac{\partial \ln \rho_B}{\partial \ln p} \right)_E + \left(\frac{1}{\rho_A} - \frac{1}{\rho_B} \right) \left(\rho_B h_B \frac{\partial \ln h_B}{\partial \ln p} - p \right)_E \bigg/ (h_A - h_B)$$

The sensitivity of the results to the, as yet, unspecified quantitives ϵ_A and ϵ_B does not emerge until one substitutes vapor and liquid for the phases A and B ($A = V$, $B = L$, and $\alpha_A = \alpha$, $\alpha_B = 1 - \alpha$ for simplicity). The functions f_L, f_B, g_L, and g_V then become

$$f_V = \left(\frac{\partial \ln \rho_V}{\partial \ln p} \right)_S ; \quad f_L = \left(\frac{\partial \ln \rho_L}{\partial \ln p} \right)_S \tag{6.23}$$

$$g_V = \left(\frac{\partial \ln \rho_V}{\partial \ln p} \right)_E + \left(1 - \frac{\rho_V}{\rho_L} \right) \left(\frac{h_L}{L} \frac{\partial \ln h_L}{\partial \ln p} + \frac{\partial \ln L}{\partial \ln p} - \frac{p}{L\rho_V} \right)_E$$

$$g_L = \left(\frac{\partial \ln \rho_L}{\partial \ln p} \right)_E + \left(\frac{\rho_L}{\rho_V} - 1 \right) \left(\frac{h_L}{L} \frac{\partial \ln h_L}{\partial \ln p} - \frac{p}{L\rho_L} \right)_E$$

where $L = h_V - h_L$ is the latent heat. It is normally adequate to approximate f_V and f_L by the reciprocal of the ratio of specific heats for the gas and zero respectively. Thus f_V is of order unity and f_L is very small. Furthermore g_L and g_V can readily be calculated for any fluid as functions of pressure or temperature. Some particular values are shown in Figure 6.2. Note that g_V is close to unity for most fluids except in the neighborhood of the critical point. On the other hand, g_L can be a large number that varies considerably with pressure. To a first approximation, g_L is given by $g^*(p_C/p)^\eta$ where p_C is the critical pressure and, as indicated in figure 6.2, g^* and η are respectively 1.67 and 0.73 for water. Thus, in summary, $f_L \approx 0$, f_V and g_V are of order unity, and g_L varies significantly with pressure and may be large.

With these magnitudes in mind, we now examine the sensitivity of $1/\rho c^2$ to the interacting fluid fractions ϵ_L and ϵ_V:

$$\frac{1}{\rho c^2} = \frac{\alpha}{p} \left[(1 - \epsilon_V)f_V + \epsilon_V g_V \right] + \frac{(1 - \alpha)}{p} \epsilon_L g_L \tag{6.24}$$

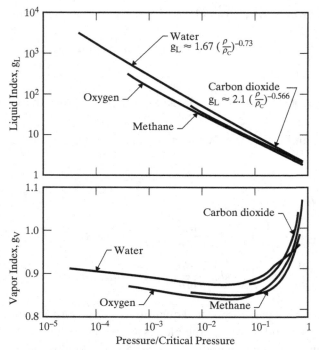

Figure 6.2 Typical values of the liquid index, g_L, and the vapor index, g_V, for various fluids.

Using $g_L = g^*(p_C/p)^\eta$ this is written for future convenience in the form:

$$\frac{1}{\rho c^2} = \frac{\alpha k_V}{p} + \frac{(1-\alpha)k_L}{p^{1+\eta}} \qquad (6.25)$$

where $k_V = (1 - \epsilon_V)f_V + \epsilon_V g_V$ and $k_L = \epsilon_L g^*(p_C)^\eta$. Note first that the result is rather insensitive to ϵ_V since f_V and g_V are both of order unity. On the other hand $1/\rho c^2$ is sensitive to the interacting liquid fraction ϵ_L though this sensitivity disappears as α approaches 1, in other words for mist flow. Thus the choice of ϵ_L is most important at low vapor volume fractions (for bubbly flows). In such cases, one possible qualitative estimate is that the interacting liquid fraction, ϵ_L, should be of the same order as the gas volume fraction, α. In Section 6.6 we will examine the effect of the choice of ϵ_L and ϵ_V on a typical vapor/liquid flow and compare the model with experimental measurements.

6.4 Barotropic Relations

Conceptually, the expressions for the sonic velocity, Equations (6.13), (6.14), (6.15), or (6.24), need only be integrated (after substituting $c^2 = dp/d\rho$) in order to obtain the barotropic relation, $p(\rho)$, for the mixture. In practice this is algebraically complicated except for some of the simpler forms for c^2.

Consider first the case of the two-component mixture in the absence of mass exchange or surface tension as given by Equation (6.14). It will initially be assumed that the gas volume fraction is not too small so that Equation (6.15) can be used; we will return later to the case of small gas volume fraction. It is also assumed that the liquid or solid density, ρ_L, is constant and that $p \propto \rho_G^k$. Furthermore it is convenient, as in gas dynamics, to choose reservoir conditions, $p = p_o, \alpha = \alpha_o$, $\rho_G = \rho_{Go}$ to establish the integration constants. Then it follows from the integration of Equation (6.15) that

$$\rho = \rho_o(1 - \alpha)/(1 - \alpha_o) \tag{6.26}$$

and that

$$\frac{p}{p_o} = \left[\frac{\alpha_o(1 - \alpha)}{(1 - \alpha_o)\alpha}\right]^k = \left[\frac{\alpha_o \rho}{\rho_o - (1 - \alpha_o)\rho}\right]^k \tag{6.27}$$

where $\rho_o = \rho_L(1 - \alpha_o) + \rho_{Go}\alpha_o$. It also follows that, written in terms of α,

$$c^2 = \frac{kp_o}{\rho_o} \frac{(1 - \alpha)^{k-1}}{\alpha^{k+1}} \frac{\alpha_o^k}{(1 - \alpha_o)^{k-1}} \tag{6.28}$$

As will be discussed later, Tangren, Dodge, and Seifert (1949) first made use of a more limited form of the barotropic relation of Equation (6.27) to evaluate the one-dimensional flow of gas/liquid mixtures in ducts and nozzles.

In the case of very small gas volume fractions, α, it may be necessary to include the liquid compressibility term, $1 - \alpha/\rho_L c_L^2$, in Equation (6.14). Exact integration then becomes very complicated. However, it is sufficiently accurate at small gas volume fractions to approximate the mixture density ρ by $\rho_L(1 - \alpha)$, and then integration (assuming $\rho_L c_L^2 = $ constant) yields

$$\frac{\alpha}{(1 - \alpha)} = \left[\frac{\alpha_o}{(1 - \alpha_o)} + \frac{k}{(k+1)} \frac{p_o}{\rho_L c_L^2}\right]\left(\frac{p_o}{p}\right)^{\frac{1}{k}} - \frac{k}{(k+1)} \frac{p_o}{\rho_L c_L^2} \frac{p}{p_o} \tag{6.29}$$

and the sonic velocity can be expressed in terms of p/p_o alone by using Equation (6.29) and noting that

$$c^2 = \frac{p}{\rho_L} \frac{\left[1 + \frac{\alpha}{(1-\alpha)}\right]^2}{\left[\frac{1}{k}\frac{\alpha}{(1-\alpha)} + \frac{p}{\rho_L c_L^2}\right]} \tag{6.30}$$

Implicit within Equation (6.29) is the barotropic relation, $p(\alpha)$, analogous to Equation (6.27). Note that Equation (6.29) reduces to Equation (6.27) when $p_o/\rho_L c_L^2$ is set equal to zero. Indeed, it is clear from Equation (6.29) that the liquid compressibility has a negligible effect only if $\alpha_o \gg p_o/\rho_L c_L^2$. This parameter, $p_o/\rho_L c_L^2$, is usually quite small. For example, for saturated water at $5 \times 10^7 \, kg/msec^2$ (500 psi) the value of $p_o/\rho_L c_L^2$ is approximately 0.03. Nevertheless, there are many practical problems in which one is concerned with the discharge of a predominantly liquid medium from high pressure containers, and under these circumstances it can be important to include the liquid compressibility effects.

Now turning attention to a two-phase rather than two-component homogeneous mixture, the particular form of the sonic velocity given in Equation (6.25) may be integrated to yield the implicit barotropic relation

$$\frac{\alpha}{1-\alpha} = \left[\frac{\alpha_o}{(1-\alpha_o)} + \frac{k_L p_o^{-\eta}}{(k_V - \eta)}\right]\left(\frac{p_o}{p}\right)^{k_V} - \left[\frac{k_L p_o^{-\eta}}{(k_V - \eta)}\right]\left(\frac{p_o}{p}\right)^{\eta} \tag{6.31}$$

in which the approximation $\rho \approx \rho_L(1-\alpha)$ has been used. As before, c^2 may be expressed in terms of p/p_o alone by noting that

$$c^2 = \frac{p}{\rho_L}\frac{\left[1 + \frac{\alpha}{1-\alpha}\right]^2}{\left[k_V\frac{\alpha}{(1-\alpha)} + k_L p^{-\eta}\right]} \tag{6.32}$$

Finally, we note that close to $\alpha = 1$ the Equations (6.31) and (6.32) may fail because the approximation $\rho \approx \rho_L(1-\alpha)$ is not sufficiently accurate.

6.5 Nozzle Flows

The barotropic relations of the last section can be used in conjunction with the steady, one-dimensional continuity and frictionless momentum equations,

$$\frac{d}{ds}(\rho A u) = 0 \tag{6.33}$$

and

$$u\frac{du}{ds} = -\frac{1}{\rho}\frac{dp}{ds} \tag{6.34}$$

to synthesize homogeneous multiphase flow in ducts and nozzles. The predicted phenomena are qualitatively similar to those in one-dimensional gas dynamics. The results for isothermal, two-component flow were first detailed by Tangren, Dodge, and Seifert (1949); more general results for any polytropic index are given in this section.

Using the barotropic relation given by Equation (6.27) and Equation (6.26) for the mixture density, ρ, to eliminate p and ρ from the momentum Equation (6.34), one obtains

$$u\,du = \frac{k p_o}{\rho_o}\frac{\alpha_o^k}{(1-\alpha_o)^{k-1}}\frac{(1-\alpha)^{k-2}}{\alpha^{k+1}}d\alpha \tag{6.35}$$

which upon integration and imposition of the reservoir condition, $u_o = 0$, yields

$$u^2 = \frac{2k p_o}{\rho_o}\frac{\alpha_o^k}{(1-\alpha_o)^{k-1}}\left[\frac{1}{k}\left\{\left(\frac{1-\alpha_o}{\alpha_o}\right)^k - \left(\frac{1-\alpha}{\alpha}\right)^k\right\} + \quad \text{either}\right.$$

$$\frac{1}{(k-1)}\left\{\left(\frac{1-\alpha_o}{\alpha_o}\right)^{k-1} - \left(\frac{1-\alpha}{\alpha}\right)^{k-1}\right\}\right] \quad \text{if } k \neq 1$$

$$\text{or} \quad ln\left\{\frac{(1-\alpha_o)\alpha}{\alpha_o(1-\alpha)}\right\}\right] \quad \text{if } k = 1 \tag{6.36}$$

Given the reservoir conditions p_o and α_o as well as the polytropic index k and the liquid density (assumed constant), this relates the velocity, u, at any position in the duct to the gas volume fraction, α, at that location. The pressure, p, density, ρ, and volume fraction, α, are related by Equations (6.26) and (6.27). The continuity equation,

$$A = \text{Constant}/\rho u = \text{Constant}/u(1 - \alpha) \qquad (6.37)$$

completes the system of equations by permitting identification of the location where p, ρ, u, and α occur from knowledge of the cross-sectional area, A.

As in gas dynamics the conditions at a throat play a particular role in determining both the overall flow and the mass flow rate. This results from the observation that Equations (6.33) and (6.31) may be combined to obtain

$$\frac{1}{A}\frac{dA}{ds} = \frac{1}{\rho}\frac{dp}{ds}\left(\frac{1}{u^2} - \frac{1}{c^2}\right) \qquad (6.38)$$

where $c^2 = dp/d\rho$. Hence at a throat where $dA/ds = 0$, either $dp/ds = 0$, which is true when the flow is entirely subsonic and unchoked; or $u = c$, which is true when the flow is choked. Denoting choked conditions at a throat by the subscript $*$, it follows by equating the right-hand sides of Equations (6.28) and (6.36) that the gas volume fraction at the throat, α_*, must be given when $k \neq 1$ by the solution of

$$\frac{(1 - \alpha_*)^{k-1}}{2\alpha_*^{k+1}} = \frac{1}{k}\left\{\left(\frac{1 - \alpha_o}{\alpha_o}\right)^k - \left(\frac{1 - \alpha_*}{\alpha_*}\right)^k\right\}$$

$$+ \frac{1}{(k-1)}\left\{\left(\frac{1 - \alpha_o}{\alpha_o}\right)^{k-1} - \left(\frac{1 - \alpha_*}{\alpha_*}\right)^{k-1}\right\} \qquad (6.39)$$

or, in the case of isothermal gas behavior $(k = 1)$, by the solution of

$$\frac{1}{2\alpha_*^2} = \frac{1}{\alpha_o} - \frac{1}{\alpha_*} + \ln\left\{\frac{(1 - \alpha_o)\alpha_*}{\alpha_o(1 - \alpha_*)}\right\} \qquad (6.40)$$

Thus the throat gas volume fraction, α_*, under choked flow conditions is a function only of the reservoir gas volume fraction, α_o, and the polytropic index. Solutions of Equations (6.39) and (6.40) for two typical cases, $k = 1.4$ and $k = 1.0$, are shown in Figure 6.3. The corresponding ratio of the choked throat pressure, p_*, to the reservior pressure, p_o, follows immediately from Equation (6.27) given $\alpha = \alpha_*$ and is also shown in Figure 6.3. Finally, the choked mass flow rate, $\dot{m}_c$, follows as $\rho_* A_* c_*$ where A_* is the cross-sectional area of the throat and

$$\frac{\dot{m}_c}{A_*(p_o\rho_o)^{\frac{1}{2}}} = k^{\frac{1}{2}}\frac{\alpha_o^{\frac{k}{2}}}{(1 - \alpha_o)^{\frac{k+1}{2}}}\left(\frac{1 - \alpha_*}{\alpha_*}\right)^{\frac{k+1}{2}} \qquad (6.41)$$

This dimensionless choked mass flow rate is exhibited in Figure 6.4 for $k = 1.4$ and $k = 1$.

Data from the experiments of Symington (1978) and Muir and Eichhorn (1963) are included in Figures 6.3 and 6.4. Symington's data on the critical pressure ratio (Figure 6.3) is in good agreement with the isothermal $(k = 1)$ analysis indicating that,

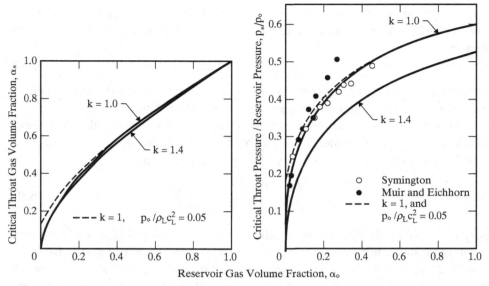

Figure 6.3 Critical or choked flow throat characteristics for the flow of a two-component gas/liquid mixture through a nozzle. On the left is the throat gas volume fraction as a function of the reservoir gas volume fraction, α_o, for gas polytropic indices of $k = 1.0$ and 1.4 and an incompressible liquid (solid lines) and for $k = 1$ and a compressible liquid with $p_o/\rho_L c_L^2 = 0.05$ (dashed line). On the right are the corresponding ratios of critical throat pressure to reservoir pressure. Also shown is the experimental data of Symington (1978) and Muir and Eichhorn (1963).

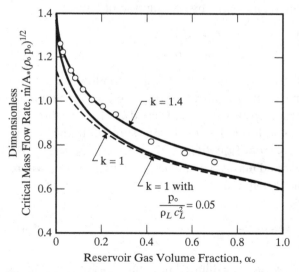

Figure 6.4 Dimensionless critical mass flow rate, $\dot{m}/A_*(p_o\rho_o)^{\frac{1}{2}}$, as a function of α_o for choked flow of a gas/liquid flow through a nozzle. Solid lines are incompressible liquid results for polytropic indices of 1.4 and 1.0. Dashed line shows effect of liquid compressibility for $p_o/\rho_L c_L^2 = 0.05$. The experimental data ($\odot$) are from Muir and Eichhorn (1963).

at least in his experiments, the heat transfer between the bubbles and the liquid is large enough to maintain constant gas temperature in the bubbles. On the other hand, the experiments of Muir and Eichhorn yielded larger critical pressure ratios and flow rates than the isothermal theory. However, Muir and Eichhorn measured significant slip between the bubbles and the liquid (strictly speaking the abscissa for their data in Figures 6.3 and 6.4 should be the upstream volumetric quality rather than the void fraction), and the discrepancy could be due to the errors introduced into the present analysis by the neglect of possible relative motion (see also van Wijngaarden 1972).

Finally, the pressure, volume fraction, and velocity elsewhere in the duct or nozzle can be related to the throat conditions and the ratio of the area, A, to the throat area, A_*. These relations, which are presented in Figures 6.5, 6.6, and 6.7 for the case $k = 1$ and various reservoir volume fractions, α_o, are most readily obtained in the following manner. Given α_o and k, p_*/p_o and α_* follow from Figure 6.3. Then for p/p_o or p/p_*, α and u follow from Equations (6.27) and (6.36) and the corresponding A/A_* follows by using Equation (6.37). The resulting charts, Figures 6.5, 6.6, and 6.7, can then be used in the same way as the corresponding graphs in gas dynamics.

If the gas volume fraction, α_o, is sufficiently small so that it is comparable with $p_o/\rho_L c_L^2$, then the barotropic Equation (6.29) should be used instead of

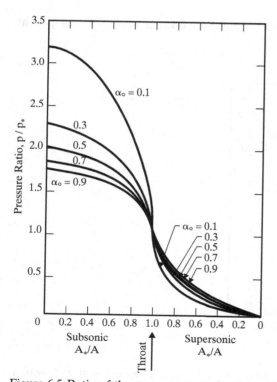

Figure 6.5 Ratio of the pressure, p, to the throat pressure, p_*, for two-component flow in a duct with isothermal gas behavior.

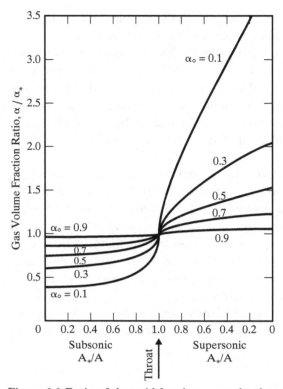

Figure 6.6 Ratio of the void fraction, α, to the throat void fraction, α_*, for two-component flow in a duct with isothermal gas behavior.

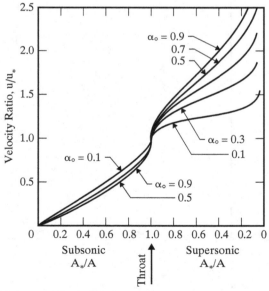

Figure 6.7 Ratio of the velocity, u, to the throat velocity, u_*, for two-component flow in a duct with isothermal gas behavior.

Equation (6.27). In cases like this in which it is sufficient to assume that $\rho \approx \rho_L(1-\alpha)$, integration of the momentum Equation (6.34) is most readily accomplished by writing it in the form

$$\frac{\rho_L}{p_o}\frac{u^2}{2} = 1 - \frac{p}{p_o} + \int_{p/p_o}^{1}\left(\frac{\alpha}{1-\alpha}\right)d\left(\frac{p}{p_o}\right) \tag{6.42}$$

Then substitution of Equation (6.29) for $\alpha/(1-\alpha)$ leads in the present case to

$$u^2 = \frac{2p_o}{\rho_L}\left[1 - \frac{p}{p_o} + \frac{k}{2(k+1)}\frac{p_o}{\rho_L c_L^2}\left\{\frac{p^2}{p_o^2} - 1\right\} + \quad \text{either}\right.$$

$$\frac{k}{(k-1)}\left\{\frac{\alpha_o}{1-\alpha_o} + \frac{k}{(k+1)}\frac{p_o}{\rho_L c_L^2}\right\}\left\{1 - \left(\frac{p}{p_o}\right)^{\frac{k-1}{k}}\right\}\right] \quad \text{for } k \neq 1$$

$$\text{or} \quad \left\{\frac{\alpha_o}{1-\alpha_o} + \frac{1}{2}\frac{p_o}{\rho_L c_L^2}\right\}\ln\left(\frac{p_o}{p}\right)\right] \quad \text{for } k = 1 \tag{6.43}$$

The throat pressure, p_* (or rather p_*/p_o), is then obtained by equating the velocity u for $p = p_*$ from Equation (6.43) to the sonic velocity c at $p = p_*$ obtained from Equation (6.30). The resulting relation, though algebraically complicated, is readily solved for the critical pressure ratio, p_*/p_o, and the throat gas volume fraction, α_*, follows from Equation (6.29). Values of p_*/p_o for $k = 1$ and $k = 1.4$ are shown in Figure 6.3 for the particular value of $p_o/\rho_L c_L^2$ of 0.05. Note that the most significant deviations caused by liquid compressibility occur for gas volume fractions of the order of 0.05 or less. The corresponding dimensionless critical mass flow rates, $\dot{m}/A_*(\rho_o p_o)^{\frac{1}{2}}$, are also readily calculated from

$$\frac{\dot{m}}{A_*(\rho_o p_o)^{\frac{1}{2}}} = \frac{(1-\alpha_*)c_*}{[p_o(1-\alpha_o)/\rho_L]^{\frac{1}{2}}} \tag{6.44}$$

and sample results are shown in Figure 6.4.

6.6 Vapor/Liquid Nozzle Flow

A barotropic relation, Equation (6.31), was constructed in Section 6.4 for the case of two-phase flow and, in particular, for vapor/liquid flow. This may be used to synthesize nozzle flows in a manner similar to the two-component analysis of the last section. Since the approximation $\rho \approx \rho_L(1-\alpha)$ was used in deriving both Equation (6.31) and Equation (6.42), we may eliminate $\alpha/(1-\alpha)$ from these equations to obtain the velocity, u, in terms of p/p_o:

$$\frac{\rho_L}{p_o}\frac{u^2}{2} = 1 - \frac{p}{p_o} + \frac{1}{(1-k_V)}\left[\frac{\alpha_o}{(1-\alpha_o)} + \frac{k_L p_o^{-\eta}}{(k_V - \eta)}\right]\left[1 - \left(\frac{p}{p_o}\right)^{1-k_V}\right]$$

$$- \frac{1}{(1-\eta)}\left[\frac{k_L p_o^{-\eta}}{(k_V - \eta)}\right]\left[1 - \left(\frac{p}{p_o}\right)^{1-\eta}\right] \tag{6.45}$$

To find the relation for the critical pressure ratio, p_*/p_o, the velocity, u, must equated with the sonic velocity, c, as given by Equation (6.32):

$$\frac{c^2}{2} = \frac{p}{\rho_L} \frac{\left[1 + \left\{\frac{\alpha_o}{1-\alpha_o} + k_L \frac{p_o^{-\eta}}{(k_V - \eta)}\right\} \left(\frac{p_o}{p}\right)^{k_V} - \left\{k_L \frac{p_o^{-\eta}}{(k_V - \eta)}\right\} \left(\frac{p_o}{p}\right)^{\eta}\right]^2}{2\left[k_V \left\{\frac{\alpha_o}{(1-\alpha_o)} + \frac{k_L p_o^{-\eta}}{(k_V - \eta)}\right\} \left(\frac{p_o}{p}\right)^{k_V} - \eta\left\{\frac{k_L p_o^{-\eta}}{(k_V - \eta)}\right\} \left(\frac{p_o}{p}\right)^{\eta}\right]}$$

(6.46)

Though algebraically complicated, the equation that results when the right-hand sides of Equations (6.45) and (6.46) are equated can readily be solved numerically to obtain the critical pressure ratio, p_*/p_o, for a given fluid and given values of α_o, the reservoir pressure and the interacting fluid fractions ϵ_L and ϵ_V (see Section 6.3). Having obtained the critical pressure ratio, the critical vapor volume fraction, α_*, follows from Equation (6.31) and the throat velocity, c_*, from Equation (6.46). Then the dimensionless choked mass flow rate follows from the same relation as given in Equation (6.44).

Sample results for the choked mass flow rate and the critical pressure ratio are shown in Figures 6.8 and 6.9. Results for both homogeneous frozen flow ($\epsilon_L = \epsilon_V = 0$) and for homogeneous equilibrium flow ($\epsilon_L = \epsilon_V = 1$) are presented; note that these results are independent of the fluid or the reservoir pressure, p_o. Also shown in the Figures are the theoretical results for various partially frozen cases for water at two different reservoir pressures. The interacting fluid fractions were chosen with the comment at the end of Section 6.3 in mind. Since ϵ_L is most important at low vapor volume fractions (i.e., for bubbly flows), it is reasonable to

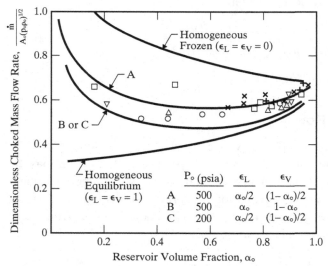

Figure 6.8 The dimensionless choked mass flow rate, $\dot{m}/A_*(p_o\rho_o)^{\frac{1}{2}}$, plotted against the reservoir vapor volume fraction, α_o, for water/steam mixtures. The data shown is from the experiments of Maneely (1962) and Neusen (1962) for $100 \rightarrow 200$ *psia* (+), $200 \rightarrow 300$ *psia* ($\times$), $300 \rightarrow 400$ *psia* ($\square$), $400 \rightarrow 500$ *psia* ($\triangle$), $500 \rightarrow 600$ *psia* ($\triangledown$) and > 600 *psia* ($*$). The theoretical lines use $g^* = 1.67$, $\eta = 0.73$, $g_V = 0.91$, and $f_V = 0.769$ for water.

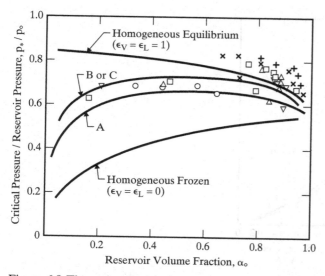

Figure 6.9 The ratio of critical pressure, p_*, to reservoir pressure, p_o, plotted against the reservoir vapor volume fraction, α_o, for water/steam mixtures. The data and the partially frozen model results are for the same conditions as in Figure 6.8.

estimate that the interacting volume of liquid surrounding each bubble will be of the same order as the bubble volume. Hence $\epsilon_L = \alpha_o$ or $\alpha_o/2$ are appropriate choices. Similarly, ϵ_V is most important at high vapor volume fractions (i.e., droplet flows), and it is reasonable to estimate that the interacting volume of vapor surrounding each droplet would be of the same order as the droplet volume; hence $\epsilon_V = (1 - \alpha_o)$ or $(1 - \alpha_o)/2$ are appropriate choices.

Figures 6.8 and 6.9 also include data obtained for water by Maneely (1962) and Neusen (1962) for various reservoir pressures and volume fractions. Note that the measured choked mass flow rates are bracketed by the homogeneous frozen and equilibrium curves and that the appropriately chosen partially frozen analysis is in close agreement with the experiments, despite the neglect (in the present model) of possible slip between the phases. The critical pressure ratio data is also in good agreement with the partially frozen analysis except for some discrepancy at the higher reservoir volume fractions.

It should be noted that the analytical approach described above is much simpler to implement than the numerical solution of the basic equations suggested by Henry and Fauske (1971). The latter does, however, have the advantage that slip between the phases was incorporated into the model.

Finally, information on the pressure, volume fraction, and velocity elsewhere in the duct (p/p_*, u/u_*, and α/α_*) as a function of the area ratio A/A_* follows from a procedure similar to that used for the noncondensable case in Section 6.5. Typical results for water with a reservoir pressure, p_o, of 500 $psia$ and using the partially frozen analysis with $\epsilon_V = \alpha_o/2$ and $\epsilon_L = (1 - \alpha_o)/2$ are presented in Figures 6.10, 6.11, and 6.12. In comparing these results with those for the two-component mixture

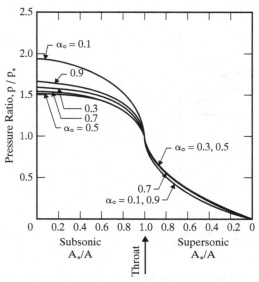

Figure 6.10 Ratio of the pressure, p, to the critical pressure, p_*, as a function of the area ratio, A_*/A, for the case of water with $g^* = 1.67$, $\eta = 0.73$, $g_V = 0.91$, and $f_V = 0.769$.

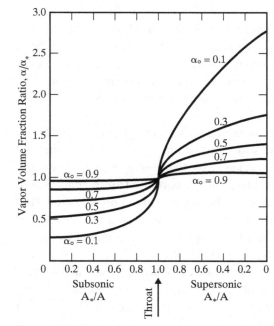

Figure 6.11 Ratio of the vapor volume fraction, α, to the critical vapor volume fraction, α_*, as a function of area ratio for the same case as Figure 6.10.

(Figures 6.5, 6.6, and 6.7) we observe that the pressure ratios are substantially smaller and do not vary monotonically with α_o. The volume fraction changes are smaller, while the velocity gradients are larger.

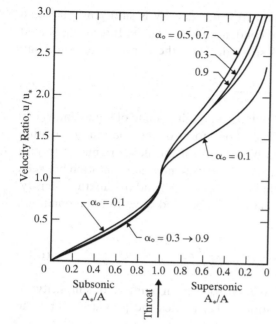

Figure 6.12 Ratio of the velocity, u, to the critical velocity, u_*, as a function of the area ratio for the same case as Figure 6.10.

6.7 Flows with Bubble Dynamics

Up to this point the analyses have been predicated on the existence of an effective barotropic relation for the homogeneous mixture. Indeed, the construction of the sonic speed in Sections 6.2 and 6.3 assumes that all the phases are in dynamic equilibrium at all times. For example, in the case of bubbles in liquids, it is assumed that the response of the bubbles to the change in pressure, δp, is an essentially instantaneous change in their volume. In practice this would only be the case if the typical frequencies experienced by the bubbles in the flow are very much smaller than the natural frequencies of the bubbles themselves (see Section 4.2). Under these circumstances the bubbles would behave quasistatically and the mixture would be barotropic.

In this section we shall examine some flows in which this criterion is not met. Then the dynamics of individual bubbles as manifest by the Rayleigh-Plesset Equation (2.12) should be incorporated into the solutions of the problem. The mixture will no longer behave barotropically.

Viewing it from another perspective, we note that analyses of cavitating flows often consist of using a single-phase liquid pressure distribution as input to the Rayleigh-Plesset equation. The result is the history of the size of individual cavitating bubbles as they progress along a streamline in the otherwise purely liquid flow. Such an approach entirely neglects the interactive effects that the cavitating bubbles have on themselves and on the pressure and velocity of the liquid flow. The analysis that follows incorporates these interactions using the equations for nonbarotropic homogeneous flow.

It is assumed that the ratio of liquid to vapor density is sufficiently large so that the volume of liquid evaporated or condensed is negligible. It is also assumed that bubbles are neither created or destroyed. Then the appropriate continuity equation is

$$\frac{\partial u_i}{\partial x_i} = \frac{\eta}{(1 + \eta\tau)} \frac{D\tau}{Dt} \tag{6.47}$$

where η is the population or number of bubbles per unit volume of liquid and $\tau(x_i, t)$ is the volume of individual bubbles. The above form of the continuity equation assumes that η is uniform; such would be the case if the flow originated from a uniform stream of uniform population and if there were no relative motion between the bubbles and the liquid. Note also that $\alpha = \eta\tau/(1 + \eta\tau)$ and the mixture density, $\rho \approx \rho_L(1 - \alpha) = \rho_L/(1 + \eta\tau)$. This last relation can be used to write the momentum Equation (6.3) in terms of τ rather than ρ:

$$\rho_L \frac{Du_i}{Dt} = -(1 + \eta\tau)\frac{\partial p}{\partial x_i} \tag{6.48}$$

The hydrostatic pressure gradient due to gravity has been omitted for simplicity.

Finally the Rayleigh-Plesset Equation (2.12) relates the pressure p and the bubble volume, $\tau = \frac{4}{3}\pi R^3$:

$$R\frac{D^2R}{Dt^2} + \frac{3}{2}\left(\frac{DR}{Dt}\right)^2 = \frac{p_B - p}{\rho_L} - \frac{2S}{\rho_L R} - \frac{4\nu_L}{R}\frac{DR}{Dt} \tag{6.49}$$

where p_B, the pressure within the bubble, will be represented by the sum of a partial pressure, p_V, of the vapor plus a partial pressure of noncondensable gas as given in Equation (2.11).

Equations (6.47), (6.48), and (6.49) can, in theory, be solved to find the unknowns $p(x_i, t)$, $u_i(x_i, t)$, and $\tau(x_i, t)$ (or $R(x_i, t)$) for any bubbly cavitating flow. In practice the nonlinearities in the Rayleigh-Plesset equation and in the Lagrangian derivative, $D/Dt = \partial/\partial t + u_i\partial/\partial x_i$, present serious difficulties for all flows except those of the simplest geometry. In the following sections several such flows are examined in order to illustrate the interactive effects of bubbles in cavitating flows and the role played by bubble dynamics in homogeneous flows.

6.8 Acoustics of Bubbly Mixtures

One class of phenomena in which bubble dynamics can play an important role is the acoustics of dilute bubbly mixtures. When the acoustic excitation frequency approaches the natural frequency of the bubbles, the latter no longer respond in the quasistatic manner assumed in Section 6.2, and both the propagation speed and the acoustic attenuation are significantly altered. An excellent review of this subject is given by van Wijngaarden (1972) and we will include here only a summary of the key results. This class of problems has the advantage that the magnitude of the perturbations is small so that the equations of the preceding section can be greatly simplified by linearization.

Hence the pressure, p, will be represented by the following sum:

$$p = \bar{p} + Re\left\{\tilde{p}e^{j\omega t}\right\} \tag{6.50}$$

where $\bar{p}$ is the mean pressure, ω is the frequency, and $\tilde{p}$ is the small amplitude pressure perturbation. The response of a bubble will be similarly represented by a perturbation, φ, to its mean radius, R_o, such that

$$R = R_o\left[1 + Re\left\{\varphi e^{j\omega t}\right\}\right] \tag{6.51}$$

and the linearization will neglect all terms of order φ^2 or higher.

The literature on the acoustics of dilute bubbly mixtures contains two complementary analytical approaches. In important papers, Foldy (1945) and Carstensen and Foldy (1947) applied the classical acoustical approach and treated the problem of multiple scattering by randomly distributed point scatterers representing the bubbles. The medium is assumed to be very dilute ($\alpha \ll 1$). The multiple scattering produces both coherent and incoherent contributions. The incoherent part is beyond the scope of this text. The coherent part, which can be represented by Equation (6.50), was found to satsify a wave equation and yields a dispersion relation for the wavenumber, k, of plane waves, which implies a phase velocity, $c_k = \omega/k$, given by (see van Wijngaarden 1972)

$$\frac{1}{c_k^2} = \frac{k^2}{\omega^2} = \frac{1}{c_L^2} + \frac{1}{c_o^2}\left[1 - \frac{j\delta_D\omega}{\omega_N} - \frac{\omega^2}{\omega_N^2}\right]^{-1} \tag{6.52}$$

Here c_L is the sonic speed in the liquid, c_o is the sonic speed arising from Equation (6.15) when $\alpha\rho_G \ll (1-\alpha)\rho_L$,

$$c_o^2 = k\bar{p}/\rho_L\alpha(1-\alpha) \tag{6.53}$$

ω_N is the natural frequency of a bubble in an infinite liquid (Section 4.2), and δ_D is a dissipation coefficient that will be discussed shortly. It follows from Equation (6.52) that scattering from the bubbles makes the wave propagation dispersive since c_k is a function of the frequency, ω.

As described by van Wijngaarden (1972) an alternative approach is to linearize the fluid mechanical Equations (6.47), (6.48), and (6.49), neglecting any terms of order φ^2 or higher. In the case of plane wave propagation in the direction x (velocity u) in a frame of reference relative to the mixture (so that the mean velocity is zero), the convective terms in the Lagrangian derivatives, D/Dt, are of order φ^2 and the three governing equations become

$$\frac{\partial u}{\partial x} = \frac{\eta}{(1+\eta\tau)}\frac{\partial \tau}{\partial t} \tag{6.54}$$

$$\rho_L\frac{\partial u}{\partial t} = -(1+\eta\tau)\frac{\partial p}{\partial x} \tag{6.55}$$

$$R\frac{\partial^2 R}{\partial t^2} + \frac{3}{2}\left(\frac{\partial R}{\partial t}\right)^2 = \frac{1}{\rho_L}\left[p_V + p_{Go}\left(\frac{R_o}{R}\right)^{3k} - p\right] - \frac{2S}{\rho_L R} - \frac{4\nu_L}{R}\frac{\partial R}{\partial t} \tag{6.56}$$

Assuming for simplicity that the liquid is incompressible (ρ_L = constant) and eliminating two of the three unknown functions from these relations, one obtains the following equation for any one of the three perturbation quantities ($q = \varphi$, $\tilde{p}$, or $\tilde{u}$, the velocity perturbation):

$$3\alpha_o(1-\alpha_o)\frac{\partial^2 q}{\partial t^2} = \left[\frac{3kp_{Go}}{\rho_L} - \frac{2S}{\rho_L R_o}\right]\frac{\partial^2 q}{\partial x^2} + R_o^2 \frac{\partial^4 q}{\partial x^2 \partial t^2} + 4\nu_L \frac{\partial^3 q}{\partial x^2 \partial t} \qquad (6.57)$$

where α_o is the mean void fraction given by $\alpha_o = \eta\tau_o/(1+\eta\tau_o)$. This equation governing the acoustic perturbations is given by van Wijngaarden, though we have added the surface tension term. Since the mean state must be in equilibrium, the mean liquid pressure, $\bar{p}$, is related to p_{Go} by

$$\bar{p} = p_V + p_{Go} - \frac{2S}{R_o} \qquad (6.58)$$

and hence the term in square brackets in Equation (6.57) may be written in the alternate forms

$$\frac{3kp_{Go}}{\rho_L} - \frac{2S}{\rho_L R_o} = \frac{3k}{\rho_L}(\bar{p}-p_V) + \frac{2S}{\rho_L R_o}(3k-1) = R_o^2 \omega_N^2 \qquad (6.59)$$

where ω_N is the natural frequency of a single bubble in an infinite liquid (see Section 4.2).

Results for the propagation of a plane wave in the positive x direction are obtained by substituting $q = e^{-jkx}$ in Equation (6.57) to produce the following dispersion relation:

$$c_k^2 = \frac{\omega^2}{k^2} = \frac{\left[\frac{3k}{\rho_L}(\bar{p}-p_V) + \frac{2S}{\rho_L R_o}(3k-1)\right] + 4j\omega\nu_L - \omega^2 R_o^2}{3\alpha_o(1-\alpha_o)} \qquad (6.60)$$

Note that at the low frequencies for which one would expect quasistatic bubble behavior ($\omega \ll \omega_N$) and in the absence of vapor ($p_V = 0$) and surface tension, this reduces to the sonic velocity given by Equation (6.15) when $\rho_G \alpha \ll \rho_L(1-\alpha)$. Furthermore, Equation (6.60) may be written as

$$c_k^2 = \frac{\omega^2}{k^2} = \frac{R_o^2 \omega_N^2}{3\alpha_o(1-\alpha_o)}\left[1 + j\frac{\delta_D\omega}{\omega_N} - \frac{\omega^2}{\omega_N^2}\right] \qquad (6.61)$$

where $\delta_D = 4\nu_L/\omega_N R_o^2$. For the incompressible liquid assumed here this is identical to Equation (6.52) obtained using the Foldy multiple scattering approach (the difference in sign for the damping term results from using $j(\omega t - kx)$ rather than $j(kx - \omega t)$ and is inconsequential).

In the above derivation, the only damping mechanism that was included was that due to viscous effects on the radial motion of the bubbles. As discussed in Section 4.4, other damping mechanisms (thermal and acoustic radiation) that may affect radial bubble motion can be included in approximate form in the above analysis by defining an "effective" damping, δ_D, or, equivalently, an effective liquid viscosity, $\mu_E = \omega_N R_o^2 \delta_D/4$.

The real and imaginary parts of k as defined by Equation (6.61) lead respectively to a sound speed and an attenuation that are both functions of the frequency of the perturbations. A number of experimental investigations have been carried out (primarily at very small α) to measure the sound speed and attenuation in bubbly gas/liquid mixtures. This data is reviewed by van Wijngaarden (1972) who concentrates on the more recent experiments of Fox, Curley, and Lawson (1955), Macpherson (1957), and Silberman (1957), in which the bubble size distribution was more accurately measured and controlled. In general, the comparison between the experimental and theoretical propagation speeds is good, as illustrated by Figure 6.13. One of the primary experimental difficulties illustrated in both Figures 6.13 and 6.14 is that the results are quite sensitive to the distribution of bubble sizes present in the mixture. This is caused by the fact that the bubble natural frequency is quite sensitive to the mean radius (see Section 4.2). Hence a distribution in the size of the bubbles yields broadening of the peaks in the data of Figures 6.13 and 6.14.

Though the propagation speed is fairly well predicted by the theory, the same cannot be said of the attenuation, and there remain a number of unanswered questions in this regard. Using Equation (6.61) the theoretical estimate of the damping coefficient, δ_D, pertinent to the experiments of Fox, Curley, and Lawson (1955) is 0.093. But a much greater value of $\delta_D = 0.5$ had to be used in order to produce

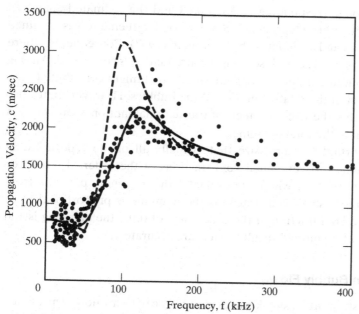

Figure 6.13 Sonic speed for water with air bubbles of mean radius, $R_o = 0.12mm$, and a void fraction, $\alpha = 0.0002$, plotted against frequency. The experimental data of Fox, Curley, and Larson (1955) is plotted along with the theoretical curve for a mixture with identical $R_o = 0.11\ mm$ bubbles (dotted line) and with the experimental distribution of sizes (solid line). These lines use $\delta = 0.5$.

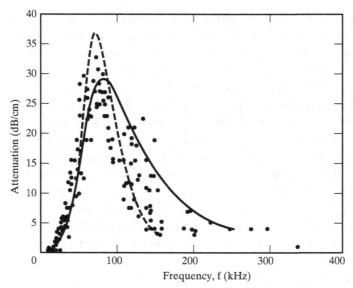

Figure 6.14 Values for the attenuation of sound waves corresponding to the sonic speed data of Figure 6.13. The attenuation in dB/cm is given by $8.69\ Im\{k\}$ where k is in cm^{-1}.

an analytical line close to the experimental data on attenuation; it is important to note that the empirical value, $\delta_D = 0.5$, has been used for the theoretical results in Figure 6.14. On the other hand, Macpherson (1957) found good agreement between a measured attenuation corresponding to $\delta_D \approx 0.08$ and the estimated analytical value of 0.079 relevant to his experiments. Similar good agreement was obtained for both the propagation and attenuation by Silberman (1957). Consequently, there appear to be some unresolved issues insofar as the attenuation is concerned. Among the effects that were omitted in the above analysis and that might contribute to the attenuation is the effect of the relative motion of the bubbles. However, Batchelor (1969) has concluded that the viscous effects of translational motion would make a negligible contribution to the total damping.

Finally, it is important to emphasize that virtually all of the reported data on attenuation is confined to very small void fractions of the order of 0.0005 or less. The reason for this is clear when one evaluates the imaginary part of k from Equation (6.61). At these small void fractions the damping is proportional to α. Consequently, at large void fraction of the order, say, of 0.05, the damping is 100 times greater and therefore more difficult to measure accurately.

6.9 Shock Waves in Bubbly Flows

The propagation and structure of shock waves in bubbly cavitating flows represent a rare circumstance in which fully nonlinear solutions of the governing equations can be obtained. Shock wave analyses of this kind have been investigated by Campbell and Pitcher (1958), Crespo (1969), Noordzij (1973), and Noordzij and van Wijngaarden (1974), among others, and for more detail the reader should consult these works. Since this chapter is confined to flows without significant relative motion, this section will not cover some of the important effects of relative motion on the structural evolution of shocks in bubbly liquids. For this the reader is referred to Noordzij and van Wijngaarden (1974).

Consider a normal shock wave in a coordinate system moving with the shock so that the flow is steady and the shock stationary. If x and u represent a coordinate and the fluid velocity normal to the shock, then continuity requires

$$\rho u = \text{constant} = \rho_1 u_1 \tag{6.62}$$

where ρ_1 and u_1 will refer to the mixture density and velocity far upstream of the shock. Hence u_1 is also the velocity of propagation of a shock into a mixture with conditions identical to those upstream of the shock. It is assumed that $\rho_1 \approx \rho_L(1-\alpha_1) = \rho_L/(1+\eta\tau_1)$ where the liquid density is considered constant and α_1, $\tau_1 = \frac{4}{3}\pi R_1^3$, and η are the void fraction, individual bubble volume, and population of the mixture far upstream.

Substituting for ρ in the equation of motion and integrating, one also obtains

$$p + \frac{\rho_1^2 u_1^2}{\rho} = \text{constant} = p_1 + \rho_1 u_1^2 \tag{6.63}$$

This expression for the pressure, p, may be substituted into the Rayleigh-Plesset equation using the observation that, for this steady flow,

$$\frac{DR}{Dt} = u\frac{dR}{dx} = u_1\frac{(1+\eta\tau)}{(1+\eta\tau_1)}\frac{dR}{dx} \tag{6.64}$$

$$\frac{D^2R}{Dt^2} = u_1^2\frac{(1+\eta\tau)}{(1+\eta\tau_1)^2}\left[(1+\eta\tau)\frac{d^2R}{dx^2} + 4\pi R^2\eta\left(\frac{dR}{dx}\right)^2\right] \tag{6.65}$$

where $\tau = \frac{4}{3}\pi R^3$ has been used for clarity. It follows that the structure of the flow is determined by solving the following equation for $R(x)$:

$$u_1^2\frac{(1+\eta\tau)^2}{(1+\eta\tau_1)^2}R\frac{d^2R}{dx^2} + \frac{3}{2}u_1^2\frac{(1+3\eta\tau)(1+\eta\tau)}{(1+\eta\tau_1)^2}\left(\frac{dR}{dx}\right)^2 \tag{6.66}$$

$$+ \frac{2S}{\rho_L R} + \frac{u_1(1+\eta\tau)}{(1+\eta\tau_1)}\frac{4\nu_L}{R}\left(\frac{dR}{dx}\right) = \frac{(p_B-p_1)}{\rho_L} + \frac{\eta(\tau-\tau_1)}{(1+\eta\tau_1)^2}u_1^2$$

It will be found that dissipation effects in the bubble dynamics (see Sections 4.3 and 4.4) strongly influence the structure of the shock. Only one dissipative term, that term due to viscous effects (last term on the left-hand side) has been included in Equation (6.66). However, note that the other dissipative effects may be incorporated approximately (see Section 4.4) by regarding ν_L as a total "effective" damping viscosity.

The pressure within the bubble is given by

$$p_B = p_V + p_{G1}\left(\tau_1/\tau\right)^k \tag{6.67}$$

and the equilibrium state far upstream must satisfy

$$p_V - p_1 + p_{G1} = 2S/R_1 \tag{6.68}$$

Furthermore, if there exists an equilibrium state far downstream of the shock (this existence will be explored shortly), then it follows from Equations (6.66) and (6.67) that the velocity, u_1, must be related to the ratio, R_2/R_1 (where R_2 is the bubble size downstream of the shock), by

$$u_1^2 = \frac{(1-\alpha_2)}{(1-\alpha_1)(\alpha_1-\alpha_2)} \left[\frac{(p_1-p_V)}{\rho_L} \left\{ \left(\frac{R_1}{R_2}\right)^{3k} - 1 \right\} \right.$$
$$\left. + \frac{2S}{\rho_L R_1} \left\{ \left(\frac{R_1}{R_2}\right)^{3k} - \frac{R_1}{R_2} \right\} \right] \tag{6.69}$$

where α_2 is the void fraction far downstream of the shock and

$$\left(\frac{R_2}{R_1}\right)^3 = \frac{\alpha_2(1-\alpha_1)}{\alpha_1(1-\alpha_2)} \tag{6.70}$$

Hence the "shock velocity," u_1, is given by the upstream flow parameters α_1, $(p_1-p_V)/\rho_L$, and $2S/\rho_L R_1$, the polytropic index, k, and the downstream void fraction, α_2. An example of the dependence of u_1 on α_1 and α_2 is shown in Figure 6.15 for selected values of $(p_1 - p_V)/\rho_L = 100\ m^2/sec^2$, $2S/\rho_L R_1 = 0.1\ m^2/sec^2$, and $k = 1.4$. Also displayed by the dotted line in this figure is the sonic velocity of the mixture, c_1, under the upstream conditions (actually the sonic velocity at zero frequency); it

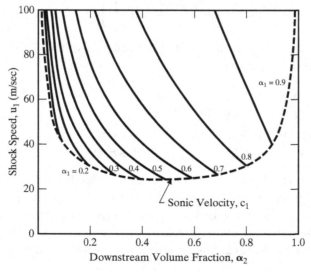

Figure 6.15 Shock speed, u_1, as a function of the upstream and downstream void fractions, α_1 and α_2, for the particular case $(p_1 - p_V)/\rho_L = 100\ m^2/sec^2$, $2S/\rho_L R_1 = 0.1\ m^2/sec^2$, and $k = 1.4$. Also shown by the dotted line is the sonic velocity, c_1, under the same upstream conditions.

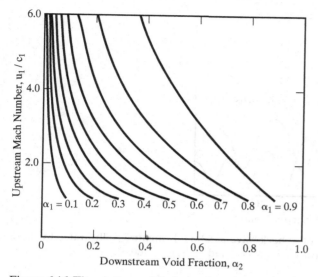

Figure 6.16 The upstream Mach number, u_1/c_1, as a function of the upstream and downstream void fractions, α_1 and α_2, for $k = 1.4$ and $R_1(p_1 - p_V)/S = 200$.

is readily shown that c_1 is given by

$$c_1^2 = \frac{1}{\alpha_1(1-\alpha_1)} \left[\frac{k(p_1 - p_V)}{\rho_L} + \left(k - \frac{1}{3} \right) \frac{2S}{\rho_L R_1} \right] \quad (6.71)$$

Alternatively, one may follow the presentation conventional in gas dynamics and plot the upstream Mach number, u_1/c_1, as a function of α_1 and α_2. The resulting graphs are functions only of two parameters, the polytropic index, k, and the parameter, $R_1(p_1 - p_V)/S$. An example is included as Figure 6.16 in which $k = 1.4$ and $R_1(p_1 - p_V)/S = 200$. It should be noted that a real shock velocity and a real sonic speed can exist even when the upstream mixture is under tension ($p_1 < p_V$). However, the numerical value of the tension, $p_V - p_1$, for which the values are real is limited to values of the parameter $R_1(p_1 - p_V)/2S > -(1 - 1/3k)$ or -0.762 for $k = 1.4$. Also note that Figure 6.16 does not change much with the parameter, $R_1(p_1 - p_V)/S$.

Bubble dynamics do not affect the results presented thus far since the speed, u_1, depends only on the equilibrium conditions upstream and downstream. However, the existence and structure of the shock depend on the bubble dynamic terms in Equation (6.66). That equation is more conveniently written in terms of a radius ratio, $r = R/R_1$, and a dimensionless coordinate, $z = x/R_1$:

$$\left(1 - \alpha_1 + \alpha_1 r^3 \right)^2 r \frac{d^2 r}{dz^2} + \frac{3}{2} \left(1 - \alpha_1 + \alpha_1 r^3 \right) \left(1 - \alpha_1 + 3\alpha_1 r^3 \right) \left(\frac{dr}{dz} \right)^2$$

$$+ \left(1 - \alpha_1 + \alpha_1 r^3 \right) \frac{4 \nu_L}{u_1 R_1} \frac{1}{r} \frac{dr}{dz} + \alpha_1 (1 - \alpha_1) \left(1 - r^3 \right)$$

$$= \frac{1}{u_1^2} \left[\frac{(p_1 - p_V)}{\rho_L} \left(r^{-3k} - 1 \right) + \frac{2S}{\rho_L R_1} \left(r^{-3k} - r^{-1} \right) \right] \quad (6.72)$$

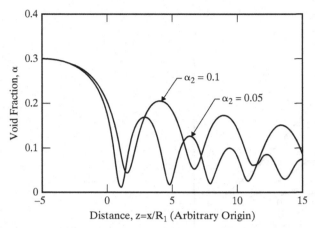

Figure 6.17 The typical structure of a shock wave in a bubbly mixture is illustrated by these examples for $\alpha_1 = 0.3$, $k = 1.4$, $R_1(p_1 - p_V)/S \gg 1$, and $u_1R_1/\nu_L = 100$.

It could also be written in terms of the void fraction, α, since

$$r^3 = \frac{\alpha}{(1-\alpha)} \frac{(1-\alpha_1)}{\alpha_1} \tag{6.73}$$

When examined in conjunction with the expression in Equation (6.69) for u_1, it is clear that the solution, $r(z)$ or $\alpha(z)$, for the structure of the shock is a function only of α_1, α_2, k, $R_1(p_1 - p_V)/S$, and the effective Reynolds number, u_1R_1/ν_L, which, as previously mentioned, should incorporate the various forms of bubble damping.

Equation (6.72) can be readily integrated numerically using Runge-Kutta procedures, and typical solutions are presented in Figure 6.17 for $\alpha_1 = 0.3$, $k = 1.4$, $R_1(p_1 - p_V)/S \gg 1$, $u_1R_1/\nu_L = 100$, and two downstream volume fractions, $\alpha_2 = 0.1$ and 0.05. These examples illustrate several important features of the structure of these shocks. First, the initial collapse is followed by many rebounds and subsequent collapses. The decay of these nonlinear oscillations is determined by the damping or u_1R_1/ν_L. Though u_1R_1/ν_L includes an effective kinematic viscosity to incorporate other contributions to the bubble damping, the value of u_1R_1/ν_L chosen for this example is probably smaller than would be relevant in many practical applications, in which we might expect the decay to be even smaller. It is also valuable to identify the nature of the solution as the damping is eliminated ($u_1R_1/\nu_L \to \infty$). In this limit the distance between collapses increases without bound until the structure consists of one collapse followed by a downstream asymptotic approach to a void fraction of α_1 (*not* α_2). In other words, no solution in which $\alpha \to \alpha_2$ exists in the absence of damping.

Another important feature in the structure of these shocks is the typical interval between the downstream oscillations. This "ringing" will, in practice, result in acoustic radiation at frequencies corresponding to this interval, and it is of importance to identify the relationship between this ring frequency and the natural frequency of the bubbles downstream of the shock. A characteristic ring frequency, ω_R, for the

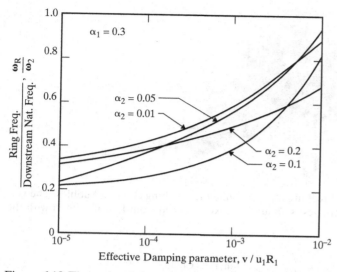

Figure 6.18 The ratio of the ring frequency downstream of a bubbly mixture shock to the natural frequency of the bubbles far downstream as a function of the effective damping parameter, ν_L/u_1R_1, for $\alpha_1 = 0.3$ and various downstream void fractions as indicated.

shock oscillations can be defined as

$$\omega_R = 2\pi u_1 / \Delta x \qquad (6.74)$$

where Δx is the distance between the first and second bubble collapses. The natural frequency of the bubbles far downstream of the shock, ω_2, is given by (see Section 4.2)

$$\omega_2^2 = \frac{3k(p_2 - p_V)}{\rho_L R_2^2} + (3k - 1)\frac{2S}{\rho_L R_2^3} \qquad (6.75)$$

and typical values for the ratio ω_R/ω_2 are presented in Figure 6.18 for $\alpha_1 = 0.3$, $k = 1.4$, $R_1(p_1 - p_V)/S \gg 1$, and various values of α_2. Similar results were obtained for quite a wide range of values of α_1. Therefore note that the frequency ratio is primarily a function of the damping and that ring frequencies up to a factor of 10 less than the natural frequency are to be expected with typical values of the damping in water. This reduction in the typical frequency associated with the collective behavior of bubbles presages the natural frequencies of bubble clouds, which are discussed in the next section.

6.10 Spherical Bubble Cloud

A second illustrative example of the effect of bubble dynamics on the behavior of a homogeneous bubbly mixture is the study of the dynamics of a finite cloud of bubbles. Clouds of bubbles occur in many circumstances. For example, breaking waves generate clouds of bubbles as illustrated in Figure 6.19, and these affect the acoustic environment in the ocean.

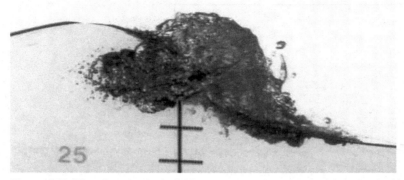

Figure 6.19 Photograph of a breaking wave showing the resulting cloud of bubbles. The vertical distances between the crosses is about 5 *cm*. Reproduced from Petroff (1993) with the author's permission.

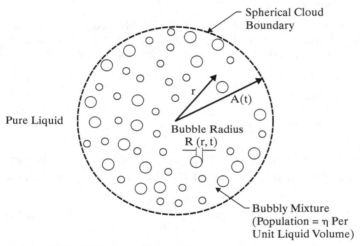

Figure 6.20 Spherical cloud of bubbles: notation.

One of the earliest investigations of the collective dynamics of bubble clouds was the work of van Wijngaarden (1964) on the oscillations of a layer of bubbles near a wall. Later d'Agostino and Brennen (1983) investigated the dynamics of a spherical cloud (see also d'Agostino and Brennen 1989, Omta 1987), and we will choose the latter as a example of that class of problems with one space dimension in which analytical solutions may be obtained but only after linearization of the Rayleigh-Plesset Equation (6.49).

The geometry of the spherical cloud is shown in Figure 6.20. Within the cloud of radius, $A(t)$, the population of bubbles per unit *liquid* volume, η, is assumed constant and uniform. The linearization assumes small perturbations of the bubbles from an equilibrium radius, R_o:

$$R(r,t) = R_o[1 + \varphi(r,t)] \ , \ |\varphi| \ll 1 \qquad (6.76)$$

We will seek the response of the cloud to a correspondingly small perturbation in the pressure at infinity, $p_\infty(t)$, which is represented by

$$p_\infty(t) = p(\infty, t) = \bar{p} + Re\left\{\tilde{p}e^{j\omega t}\right\} \qquad (6.77)$$

where $\bar{p}$ is the mean, uniform pressure and $\tilde{p}$ and ω are the perturbation amplitude and frequency, respectively. The solution will relate the pressure, $p(r,t)$, radial velocity, $u(r,t)$, void fraction, $\alpha(r,t)$, and bubble perturbation, $\varphi(r,t)$, to $\tilde{p}$. Since the analysis is linear, the response to excitation involving multiple frequencies can be obtained by Fourier synthesis.

One further restriction is necessary in order to linearize the governing Equations (6.47), (6.48), and (6.49). It is assumed that the mean void fraction in the cloud, α_o, is small so that the term $(1 + \eta\tau)$ in Equations (6.47) and (6.48) is approximately unity. Then these equations become

$$\frac{1}{r^2}\frac{\partial}{\partial r}\left(r^2 u\right) = \eta\frac{D\tau}{Dt} \qquad (6.78)$$

$$\frac{Du}{Dt} = \frac{\partial u}{\partial t} + u\frac{\partial u}{\partial r} = -\frac{1}{\rho}\frac{\partial p}{\partial r} \qquad (6.79)$$

It is readily shown that the velocity u is of order φ and hence the convective component of the material derivative is of order φ^2; thus the linearization implies replacing D/Dt by $\partial/\partial t$. It then follows from the Rayleigh-Plesset equation that to order φ

$$p(r,t) = \bar{p} - \rho R_o^2\left[\frac{\partial^2\varphi}{\partial t^2} + \omega_N^2\varphi\right]; \quad r < A(t) \qquad (6.80)$$

where ω_N is the natural frequency of an individual bubble if it were alone in an infinite fluid (equation 4.8). It must be assumed that the bubbles are in stable equilibrium in the mean state so that ω_N is real.

Upon substitution of Equations (6.76) and (6.80) into (6.78) and (6.79) and elimination of $u(r,t)$ one obtains the following equation for $\varphi(r,t)$ in the domain $r < A(t)$:

$$\frac{1}{r^2}\frac{\partial}{\partial r}\left[r^2\frac{\partial}{\partial r}\left\{\frac{\partial^2\varphi}{\partial t^2} + \omega_N^2\varphi\right\}\right] - 4\pi\eta R_o\frac{\partial^2\varphi}{\partial t^2} = 0 \qquad (6.81)$$

The incompressible liquid flow outside the cloud, $r \geq A(t)$, must have the standard solution of the form:

$$u(r,t) = \frac{C(t)}{r^2}; \quad r \geq A(t) \qquad (6.82)$$

$$p(r,t) = p_\infty(t) + \frac{\rho}{r}\frac{dC(t)}{dt} - \frac{\rho C^2}{2r^4}; \quad r \geq A(t) \qquad (6.83)$$

where $C(t)$ is of perturbation order. It follows that, to the first order in $\varphi(r,t)$, the continuity of $u(r,t)$ and $p(r,t)$ at the interface between the cloud and the pure liquid leads to the following boundary condition for $\varphi(r,t)$:

$$\left(1 + A_o\frac{\partial}{\partial r}\right)\left[\frac{\partial^2\varphi}{\partial t^2} + \omega_N^2\varphi\right]_{r=A_o} = \frac{\bar{p} - p_\infty(t)}{R_o^2\rho} \qquad (6.84)$$

The solution of Equation (6.81) under the above boundary condition is

$$\varphi(r,t) = -\frac{1}{\rho R_o^2} Re \left\{ \frac{\tilde{p}}{\omega_N^2 - \omega^2} \frac{e^{j\omega t}}{\cos \lambda A_o} \frac{\sin \lambda r}{\lambda r} \right\} ; \quad r < A_o \tag{6.85}$$

where:

$$\lambda^2 = 4\pi \eta R_o \frac{\omega^2}{\omega_N^2 - \omega^2} \tag{6.86}$$

Another possible solution involving $(\cos \lambda r)/\lambda r$ has been eliminated since $\varphi(r,t)$ must clearly be finite as $r \to 0$. Therefore in the domain $r < A_o$:

$$R(r,t) = R_o - \frac{1}{\rho R_o} Re \left\{ \frac{\tilde{p}}{\omega_N^2 - \omega^2} \frac{e^{j\omega t}}{\cos \lambda A_o} \frac{\sin \lambda r}{\lambda r} \right\} \tag{6.87}$$

$$u(r,t) = \frac{1}{\rho} Re \left\{ j \frac{\tilde{p}}{\omega} \frac{1}{r} \left(\frac{\sin \lambda r}{\lambda r} - \cos \lambda r \right) \frac{e^{j\omega t}}{\cos \lambda A_o} \right\} \tag{6.88}$$

$$p(r,t) = \bar{p} - Re \left\{ \tilde{p} \frac{\sin \lambda r}{\lambda r} \frac{e^{j\omega t}}{\cos \lambda A_o} \right\} \tag{6.89}$$

The entire flow has thus been determined in terms of the prescribed quantities A_o, R_o, η, ω, and $\tilde{p}$.

Note first that the cloud has a number of natural frequencies and modes of oscillation. From Equation (6.85) it follows that, if $\tilde{p}$ were zero, oscillations would only occur if

$$\omega = \omega_N \quad \text{or} \quad \lambda A_o = (2n - 1)\frac{\pi}{2}, \; n = 0, \pm 2 \ldots \tag{6.90}$$

and, therefore, using Equation (6.86) for λ, the natural frequencies, ω_n, of the cloud are found to be:

1. $\omega_\infty = \omega_N$, the natural frequency of an individual bubble in an infinite liquid, and
2. $\omega_n = \omega_N \left[1 + 16\eta R_o A_o^2 / \pi (2n-1)^2 \right]^{\frac{1}{2}}$; $n = 1, 2, \ldots$, which is an infinite series of frequencies of which ω_1 is the lowest. The higher frequencies approach ω_N as n tends to infinity.

The lowest natural frequency, ω_1, can be written in terms of the mean void fraction, $\alpha_o = \eta \tau_o / (1 + \eta \tau_o)$, as

$$\omega_1 = \omega_N \left[1 + \frac{4}{3\pi^2} \frac{A_o^2}{R_o^2} \frac{\alpha_o}{1 - \alpha_o} \right]^{-\frac{1}{2}} \tag{6.91}$$

Hence, the natural frequencies of the cloud will extend to frequencies much smaller than the individual bubble frequency, ω_N, if the initial void fraction, α_o, is much larger than the square of the ratio of bubble size to cloud size ($\alpha_o \gg R_o^2/A_o^2$). If the reverse is the case ($\alpha_o \ll R_o^2/A_o^2$), all the natural frequencies of the cloud are contained in a small range just below ω_N.

Typical natural modes of oscillation of the cloud are depicted in Figure 6.21, where normalized amplitudes of the bubble radius and pressure fluctuations are

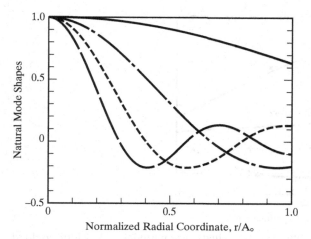

Figure 6.21 Natural mode shapes as a function of the normalized radial position, r/A_o, in the cloud for various orders $n = 1$ (solid line), 2 (dash-dotted line), 3 (dotted line), 4 (broken line). The arbitrary vertical scale represents the amplitude of the normalized undamped oscillations of the bubble radius, the pressure, and the bubble concentration per unit liquid volume. The oscillation of the velocity is proportional to the slope of these curves.

shown as functions of position, r/A_o, within the cloud. The amplitude of the radial velocity oscillation is proportional to the slope of these curves. Since each bubble is supposed to react to a uniform far field pressure, the validity of the model is limited to wave numbers, n, such that $n \ll A_o/R_o$. Note that the first mode involves almost uniform oscillations of the bubbles at all radial positions within the cloud. Higher modes involve amplitudes of oscillation near the center of the cloud, which become larger and larger relative to the amplitudes in the rest of the cloud. In effect, an outer shell of bubbles essentially shields the exterior fluid from the oscillations of the bubbles in the central core, with the result that the pressure oscillations in the exterior fluid are of smaller amplitude for the higher modes. The corresponding shielding effects during forced excitation are illustrated in Figure 6.22, which shows the distribution of the amplitude of bubble radius oscillation, $|\varphi|$, within the cloud at various excitation frequencies, ω. Note that, while the entire cloud responds in a fairly uniform manner for $\omega < \omega_N$, only a surface layer of bubbles exhibits significant response when $\omega > \omega_N$. In the latter case the entire core of the cloud is essentially shielded by the outer layer.

The variations in the response at different frequencies are shown in more detail in Figure 6.23, in which the amplitude at the cloud surface, $|\varphi(A_o,t)|$, is presented as a function of ω. The solid line corresponds to the above analysis, which did not include any bubble damping. Consequently, there are asymptotes to infinity at each of the cloud natural frequencies; for clarity we have omitted the numerous asymptotes that occur just below the bubble natural frequency, ω_N. Also shown in this figure are the corresponding results when a reasonable estimate of the damping is included in the analysis (d'Agostino and Brennen 1989). The attenuation due to the damping is much greater at the higher frequencies so that, when damping is

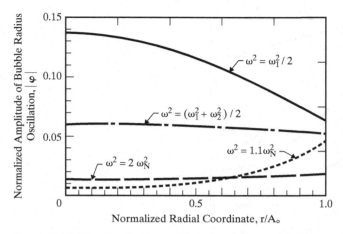

Figure 6.22 The distribution of bubble radius oscillation amplitudes, $|\varphi|$, within a cloud subjected to forced excitation at various frequencies, ω, as indicated (for the case of $\alpha_o(1-\alpha_o)A_o^2/R_o^2 = 0.822$). From d'Agostino and Brennen (1989).

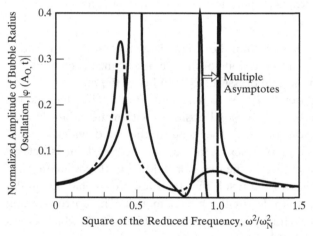

Figure 6.23 The amplitude of the bubble radius oscillation at the cloud surface, $|\varphi(A_o,t)|$, as a function of frequency (for the case of $\alpha_o(1-\alpha_o)A_o^2/R_o^2 = 0.822$). Solid line is without damping; broken line includes damping. From d'Agostino and Brennen (1989).

included (Figure 6.23), the dominant feature of the response is the lowest natural frequency of the cloud. The response at the bubble natural frequency becomes much less significant.

The effect of varying the parameter, $\alpha_o(1-\alpha_o)A_o^2/R_o^2$, is shown in Figure 6.24. Note that increasing the void fraction causes a reduction in both the amplitude and frequency of the dominant response at the lowest natural frequency of the cloud. d'Agostino and Brennen (1988) have also calculated the acoustical absorption and scattering cross-sections of the cloud that this analysis implies. Not surprisingly, the dominant peaks in the cross-sections occur at the lowest cloud natural frequency.

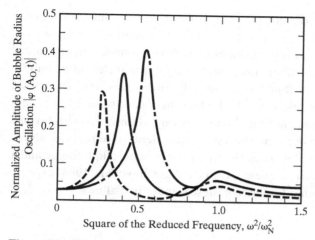

Figure 6.24 The amplitude of the bubble radius oscillation at the cloud surface, $|\varphi(A_O,t)|$, as a function of frequency for damped oscillations at three values of $\alpha_o(1-\alpha_o)A_o^2/R_o^2$ equal to 0.822 (solid line), 0.411 (dot-dash line), and 1.65 (dashed line). From d'Agostino and Brennen (1989).

It is important to emphasize that the analysis presented above is purely linear and that there are likely to be very significant nonlinear effects that may have a major effect on the dynamics and acoustics of real bubble clouds. Hanson et al. (1981) and Mørch (1980, 1981) visualize that the collapse of a cloud of bubbles involves the formation and inward propagation of a shock wave and that the focusing of this shock at the center of the cloud creates the enhancement of the noise and damage potential associated with cloud collapse (see Section 3.7). The deformations of the individual bubbles within a collapsing cloud have been examined numerically by Chahine and Duraiswami (1992), who showed that the bubbles on the periphery of the cloud develop inwardly directed reentrant jets (see Section 3.5).

Numerical investigations of the nonlinear dynamics of cavity clouds have been carried out by Chahine (1982), Omta (1987), and Kumar and Brennen (1991, 1992, 1993). Kumar and Brennen have obtained weakly nonlinear solutions to a number of cloud problems by retaining only the terms that are quadratic in the amplitude; this analysis is a natural extension of the weakly nonlinear solutions for a single bubble described in Section 4.6. One interesting phenomenon that emerges from this nonlinear analysis involves the interactions between the bubbles of different size that would commonly occur in any real cloud. The phenomenon, called "harmonic cascading" (Kumar and Brennen 1992), occurs when a relatively small number of larger bubbles begins to respond nonlinearly to some excitation. Then the higher harmonics produced will excite the much larger number of smaller bubbles at their natural frequency. The process can then be repeated to even smaller bubbles. In essence, this nonlinear effect causes a cascading of fluctuation energy to smaller bubbles and higher frequencies.

In all of the above we have focused, explicitly or implicitly, on spherical bubble clouds. Solutions of the basic equations for other, more complex geometries are not readily obtained. However, d'Agostino et al. (1988) have examined some of the characteristics of this class of flows past slender bodies (for example, the flow over a wavy surface). Clearly, in the absence of bubble dynamics, one would encounter two types of flow: subsonic and supersonic. Interestingly, the inclusion of bubble dynamics leads to three types of flow. At sufficiently low speeds one obtains the usual elliptic equations of subsonic flow. And when the sonic speed is exceeded, the equations become hyberbolic and the flow supersonic. However, with further increase in speed, the time rate of change becomes equivalent to frequencies above the natural frequency of the bubbles. Then the equations become elliptic again and a new flow regime, termed "super-resonant," occurs. d'Agostino et al. (1988) explore the consequences of this and other features of these slender body flows.

References

Batchelor, G.K. (1969). In *Fluid Dynamics Transactions*, 4, (eds: W.Fizdon, P.Kucharczyk, and W.J.Prosnak). Polish Sci. Publ., Warsaw.

Campbell, I.J. and Pitcher. A.S. (1958). Shock waves in a liquid containing gas bubbles. *Proc. Roy. Soc. London, A*, **243**, 534–545.

Carstensen, E.L. and Foldy, L.L. (1947). Propagation of sound through a liquid containing bubbles. *J. Acoust. Soc. Amer.*, **19**, 481–501.

Chahine, G.L. (1982). Cloud cavitation: theory. *Proc. 14th ONR Symp. on Naval Hydrodynamics*, 165–194.

Chahine, G.L. and Duraiswami, R. (1992). Dynamical interactions in a multibubble cloud. *ASME J. Fluids Eng.*, **114**, 680–686.

Chapman, R.B. and Plesset, M.S. (1971). Thermal effects in the free oscillation of gas bubbles. *ASME J. Basic Eng.*, **93**, 373–376.

Crespo, A. (1969). Sound and shock waves in liquids containing bubbles. *Phys. Fluids*, **12**, 2274–2282.

d'Agostino, L., and Brennen, C.E. (1983). On the acoustical dynamics of bubble clouds. *ASME Cavitation and Multiphase Flow Forum*, 72–75.

d'Agostino, L., and Brennen, C.E. (1988). Acoustical absorption and scattering cross-sections of spherical bubble clouds. *J. Acoust. Soc. Am.*, **84**, 2126–2134.

d'Agostino, L., and Brennen, C.E. (1989). Linearized dynamics of spherical bubble clouds. *J. Fluid Mech.*, **199**, 155–176.

d'Agostino, L., Brennen, C.E., and Acosta, A.J. (1988). Linearized dynamics of two-dimensional bubbly and cavitating flows over slender surfaces. *J. Fluid Mech.*, **192**, 485–509.

Foldy, L.L. (1945). The multiple scattering of waves. *Phys. Rev.*, **67**, 107–119.

Fox, F.E., Curley, S.R., and Larson, G.S. (1955). Phase velocity and absorption measurements in water containing air bubbles. *J. Acoust. Soc. Am.*, **27**, 534–539.

Gouse, S.W. and Brown, G.A. (1964). A survey of the velocity of sound in two-phase mixtures. *ASME Paper 64-WA/FE-35*.

Hanson, I., Kedrinskii, V.K., and Mørch, K.A. (1981). On the dynamics of cavity clusters. *J. Appl. Phys.*, **15**, 1725–1734.

Henry, R.E. and Fauske, H.K. (1971). The two-phase critical flow of one-component mixtures in nozzles, orifices, and short tubes. *ASME J. Heat Transfer*, **93**, 179–187.

Karplus, H.B. (1958). The velocity of sound in a liquid containing gas bubbles. *Illinois Inst. Tech. Rep. COO-248*.

Kumar, S. and Brennen, C.E. (1991). Non-linear effects in the dynamics of clouds of bubbles. *J. Acoust. Soc. Am.*, **89**, 707–714.

Kumar, S. and Brennen, C.E. (1992). Harmonic cascading in bubble clouds. *Proc. Int. Symp. on Propulsors and Cavitation, Hamburg*, 171–179.

Kumar, S. and Brennen, C.E. (1993). Some nonlinear interactive effects in bubbly cavitation clouds. *J. Fluid Mech.*, **253**, 565–591.

Macpherson, J.D. (1957). The effect of gas bubbles on sound propagation in water. *Proc. Phys. Soc. London*, **70B**, 85–92.

Maneely, D.J. (1962). A study of the expansion process of low quality steam through a de Laval nozzle. *Univ. of Calif. Radiation Lab. Rep. UCRL-6230.*

Minnaert, M. (1933). Musical air bubbles and the sound of running water. *Phil. Mag.*, **16**, 235–248.

Mørch, K.A. (1980). On the collapse of cavity cluster in flow cavitation. *Proc. First Int. Conf. on Cavitation and Inhomogenieties in Underwater Acoustics, Springer Series in Electrophysics*, **4**, 95–100.

Mørch, K.A. (1981). Cavity cluster dynamics and cavitation erosion. *Proc. ASME Cavitation and Polyphase Flow Forum*, 1–10.

Muir, T.F. and Eichhorn, R. (1963). Compressible flow of an air-water mixture through a vertical two-dimensional converging-diverging nozzle. Proc. 1963 Heat Transfer and Fluid Mechanics Institute, Stanford Univ. Press, 183–204.

Neusen, K.F. (1962). Optimizing of flow parameters for the expansion of very low quality steam. *Univ. of Calif. Radiation Lab. Rep. UCRL-6152.*

Noordzij, L. (1973). Shock waves in mixtures of liquid and air bubbles. *Ph.D. Thesis, Technische Hogeschool, Twente, Netherlands.*

Noordzij, L. and van Wijngaarden, L. (1974). Relaxation effects, caused by relative motion, on short waves in gas-bubble/liquid mixtures. *J. Fluid Mech.*, **66**, 115–143.

Omta, R. (1987). Oscillations of a cloud of bubbles of small and not so small amplitude. *J. Acoust. Soc. Am.*, **82**, 1018–1033.

Petroff, C. (1993). The interaction of breaking solitary waves with an armored bed. *Ph.D. Thesis, Calif. Inst. of Tech.*

Silberman, E. (1957). Sound velocity and attenuation in bubbly mixtures measured in standing wave tubes. *J. Acoust. Soc. Am.*, **18**, 925–933.

Symington, W.A. (1978). Analytical studies of steady and non-steady motion of a bubbly liquid. *Ph.D. Thesis, Calif. Inst. of Tech.*

Tangren, R.F., Dodge, C.H., and Seifert, H.S. (1949). Compressibility effects in two-phase flow. *J. Appl. Phys.*, **20**, No. 7, 637–645.

van Wijngaarden, L. (1964). On the collective collapse of a large number of gas bubbles in water. *Proc. 11th Int. Cong. Appl. Mech.*, Springer-Verlag, Berlin, 854–861.

van Wijngaarden, L. (1972). One-dimensional flow of liquids containing small gas bubbles. *Ann. Rev. Fluid Mech.*, **4**, 369–396.

7 Cavitating Flows

7.1 Introduction

We begin this discussion of cavitation in flows by describing the effect of the flow on a single cavitation "event." This is the term used in referring to the processes that occur when a single cavitation nucleus is convected into a region of low pressure within the flow, grows explosively to macroscopic size, and collapses when it is convected back into a region of higher pressure. Pioneering observations of individual cavitation events were made by Knapp and his associates at the California Institute of Technology in the 1940s (see, for example, Knapp and Hollander 1948) using high-speed movie cameras capable of 20,000 frames per second. Shortly thereafter Plesset (1948), Parkin (1952), and others began to model these observations of the growth and collapse of traveling cavitation bubbles using modifications of Rayleigh's original equation of motion for a spherical bubble. Many analyses and experiments on traveling bubble cavitation followed, and a brief description these is included in the next section. All of the models are based on two assumptions: that the bubbles remain spherical and that events do not interact with one another.

However, observations of real flows demonstrate that even single cavitation bubbles are often far from spherical. Indeed, they may not even be single bubbles but rather a cloud of smaller bubbles. Departure from sphericity is often the result of the interaction of the bubble with the pressure gradients and shear forces in the flow or the interaction with a solid surface. In Section 7.3 we describe some of these effects while still assuming that the events are sufficiently far apart in space and time that they do not interact with one another or modify the global liquid flow in any significant way. Often the words "limited cavitation" are used to distinguish these circumstances from the more complex phenomena that occur at higher event densities.

When the frequency of cavitation events increases in space or time such that they begin to interact with one another, a whole new set of phenomena may be manifest. They may begin to interact hydrodynamically, and some of the resulting phenomena are described and analysed in Chapter 6. Often these interaction phenomena can have important practical consequences as is the case, for example, with cloud cavitation (see section 3.7).

But increase in the density of events also causes the formation of large-scale cavitation structures either because of the coalescence of individual bubbles (often because they accumulate in regions of recirculating flow) or because a large region of the flow vaporizes. Typical large-scale structures include cavitating vortices and attached cavities. As a result, cavitating flows can exhibit a number of different kinds of cavitation; later in this chapter we shall describe some of the forms that large-scale cavitation structures can take. Some of the analytical methods used to understand and predict these structures are discussed in the next chapter.

7.2 Traveling Bubble Cavitation

Since the early work by Plesset (1948) had demonstrated some approximate validity for models of cavitation events that use the equation we now refer to as the Rayleigh-Plesset equation, Parkin (1952) was motivated to attempt a more detailed model for the growth of traveling cavitation bubbles in the flow around a body. It was assumed that the bubbles began as micron-sized nuclei in the liquid of the oncoming stream and that the bubble moved with the liquid velocity along a stream-line close to the solid surface. Cavitation inception was deemed to occur when the bubbles reached an observable size of the order of 1 *mm*. Parkin believed the lack of agreement between this theory and the experimental observations was due to the neglect of the boundary layer. Subsequent experiments by Kermeen, McGraw, and Parkin (1955) revealed that cavitation could result either from free stream nuclei as earlier assumed or from nuclei originating from imperfections in the head-form surface, which would detach when they reached a critical size. Later, Arakeri and Acosta (1973) observed that, if separation occurs close to the low-pressure region, then free stream nuclei could not only be supplied to the cavitating zone by the oncoming stream but could also be supplied by the recirculating flow down-stream of separation. Under such circumstances some of these recirculating nuclei could be remnants from a cavitation event itself, and hence there exists the possibility of hysteretic effects. Though the supply of nuclei either from the surface or from downstream may occasionally be important, the majority of the experimental observations indicate that the primary supply is from nuclei present in the incident free stream. Other viscous boundary layer effects on cavitation inception and on traveling bubble cavitation are reviewed by Holl (1969) and Arakeri (1979).

Rayleigh-Plesset models of traveling bubble cavitation that attempted to incorporate the effects of the boundary layer include the work of Oshima (1961) and Van der Walle (1962). Holl and Kornhauser (1970) added the thermal effects on bubble growth and explored the influence of initial conditions such as the size and location of the nucleus. Like Parkin's (1952) original model these improved versions continued to assume that the nucleus or bubble moves along a streamline with the fluid velocity. However, Johnson and Hsieh (1966) showed that since the streamlines that encounter the low-pressure region are close to the surface and, therefore, close to the stagnation streamline, nuclei will experience large fluid accelerations and pressure gradients as they pass close to the front stagnation point. The effect is to force the nuclei to move outwards away from the stagnation streamline. Moreover, the larger nuclei, which are those most likely to cavitate, will be displaced more than the smaller nuclei. Johnson and Hsieh termed this the "screening" effect, and more recent studies have confirmed its importance in cavitation inception. But this screening effect is only one of the effects that the accelerations and pressure gradients in the flow can have on the nucleus and on the growing and collapsing cavitation bubble. In the next section we turn to a description of these interactions.

7.3 Bubble/Flow Interactions

The maximum-modulus theorem states that maxima of a harmonic function must occur on the boundary and not in the interior of the region of solution of that function (see, for example, Titchmarsh 1947). Consequently, a pressure minimum in a steady, inviscid, potential flow must occur on the boundary of that flow (see Kirchhoff 1869, Birkhoff and Zarantonello 1957). Moreover, real fluid effects in many flows do not alter the fact that the minimum pressure occurs at or close to a solid surface. Perhaps the most common exception to this rule is in vortex cavitation, where the unsteady effects and/or viscous effects associated with vortex shedding or turbulence cause deviation from the maximum-modulus theorem; but discussion of this type of cavitation is delayed until later. In the many flows in which the minimum pressure does occur on a boundary, it follows that the cavitation bubbles that form in the vicinity of that point are likely to be affected by and to interact with that boundary, which we will assume is a solid surface. We observe, furthermore, that any curvature of the solid surface or, more specifically, of the streamlines in the vicinity of the minimum pressure point will cause pressure gradients normal to the surface, which are often substantially larger than those in the streamwise direction. These normal pressure gradients will force the bubble toward the surface and may cause substantial departure from sphericity. Consequently, even before boundary layer effects are factored into the picture, it is evident that the dynamics of individual cavitation bubbles may be significantly altered by interactions with the nearby solid surface and the flow near that surface. In this section we focus attention on these bubble/wall or bubble/flow interactions (grouped together in the term bubble/flow interactions).

Before describing some of the experimental observations of bubble/flow interactions, it is valuable to consider the relative sizes of the cavitation bubbles and the viscous boundary layer. In the flow of a uniform stream of velocity, U, around an object such as a hydrofoil with typical dimension, ℓ, the thickness of the laminar boundary layer near the minimum pressure point will be given qualitatively by $\delta = (\nu_L \ell / U)^{\frac{1}{2}}$. Parenthetically, we note that transition to turbulence usually occurs downstream of the point of minimum pressure, and consequently the appropriate boundary layer thickness for limited cavitation confined to the immediate neighborhood of the low-pressure region is the laminar boundary layer thickness. Moreover, the approximate analysis of Section 2.5 yields a typical maximum bubble radius, R_M, given by

$$R_M \approx 2\ell(-\sigma - C_{pmin}) \tag{7.1}$$

It follows that the ratio of the boundary layer thickness to the maximum bubble radius, δ/R_M, is roughly given by

$$\frac{\delta}{R_M} = \frac{1}{2(-\sigma - C_{pmin})} \left\{ \frac{\nu_L}{\ell U} \right\}^{\frac{1}{2}} \tag{7.2}$$

Therefore, provided $(-\sigma - C_{pmin})$ is of the order of 0.1 or greater, it follows that for the high Reynolds numbers, $U\ell/\nu_L$, which are typical of most of the flows in which cavitation is a problem, the boundary layer is usually much thinner than the typical dimension of the bubble. This does not mean the boundary layer is unimportant. But we can anticipate that those parts of the cavitation bubble farthest from the solid surface will interact with the primarily inviscid flow outside the boundary layer, while those parts close to the solid surface will be affected by the boundary layer.

7.4 Experimental Observations

Some of the early (and classic) observations of individual traveling cavitation bubbles by Knapp and Hollander (1948), Parkin (1952), and Ellis (1952) make mention of the deformation of the bubbles by the flow. But the focus of attention soon shifted to the easier observations of the dynamics of individual bubbles in quiescent liquid, and it is only recently that investigations of the deformation caused by the flow have resumed. Both Knapp and Hollander (1948) and Parkin (1952) observed that almost all cavitation bubbles are closer to hemispherical than spherical and that they appear to be separated from the solid surface by a thin film of liquid. Such bubbles are clearly evident in other photographs of traveling cavitation bubbles on a hydrofoil such as those of Blake et al. (1977) or Briançon-Marjollet et al. (1990).

A number of recent research efforts have focused on these bubble/flow interactions, including the work of van der Meulen and van Renesse (1989) and Briançon-Marjollet et al. (1990). Recently, Ceccio and Brennen (1991) and Kuhn de Chizelle et al. (1992a,b) have made an extended series of observations of cavitation bubbles in the flow around axisymmetric bodies, including studies of the scaling of the phenomena. Two axisymmetric body shapes were used, both of which have been employed in previous cavitation investigations. The first of these was a so-called "Schiebe body" (Schiebe 1972) which is one of a series based on the solutions for the potential flow generated by a normal source disk (Weinstein 1948) and first suggested for use in cavitation experiments by Van Tuyl (1950). One of the important characteristics of this shape is that the boundary layer does not separate in the region of low pressure within which cavitation bubbles occur. The second body had the ITTC headform shape originally used by Lindgren and Johnsson (1966) for the comparative experiments described in Section 1.15. This headform exhibits laminar separation within the region in which the cavitation bubbles occur. For both headforms, the isobars in the neighborhood of the minimum pressure point exhibit a

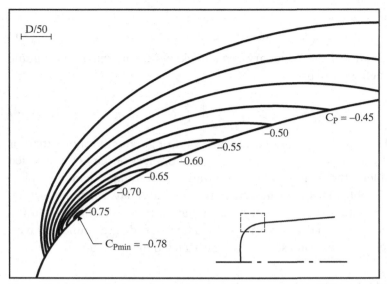

Figure 7.1 Isobars in the vicinity of the minimum pressure point on the axisymmetric Schiebe headform with values of the pressure coefficient, C_p, as indicated. The pressures were obtained from a potential flow calculation. The insert shows the headform shape and the area that has been enlarged in the main figure (dashed lines). From Schiebe (1972) and Kuhn de Chizelle et al. (1992b).

large pressure gradient normal to the surface, as illustrated by the isobars for the Schiebe body shown in Figure 7.1. This pressure gradient is associated with the curvature of the body and therefore the streamlines in the vicinity of the minimum pressure point. Consequently, at a given cavitation number, σ, the region below the vapor pressure that is enclosed between the solid surface and the $C_p = -\sigma$ isobaric surface is long and thin compared with the size of the headform. Only nuclei that pass through this thin volume will cavitate.

The observations of Ceccio and Brennen (1991) at lower Reynolds numbers will be described first. Typical photographs of bubbles on the 5.08 *cm* diameter Schiebe headform during the cycle of bubble growth and collapse are shown in Figure 7.2. Simultaneous profile and plan views provide a more complete picture of the bubble geometry. In all cases the shape during the initial growth phase was that of a spherical cap, the bubble being separated from the wall by a thin layer of liquid of the same order of magnitude as the boundary layer thickness. Later developments depend on the geometry of the headform and the Reynolds number, so we begin with the simplest case, that of the Schiebe body at relatively low Reynolds number. Typical photographs for this case are included in Figure 7.2. As the bubble begins to enter the region of adverse pressure gradient, the exterior frontal surface begins to be pushed inward, causing the profile of the bubble to appear wedge-like. Thus the collapse is initiated on the exterior frontal surface of the bubble, and this often leads to the bubble fissioning into forward and aft bubbles as seen in Figure 7.2.

Figure 7.2 A series of photographs illustrating the growth and collapse of traveling cavitation bubbles in a flow around a 5.08 *cm* diameter Schiebe headform at $\sigma = 0.45$ and a speed of 9 *m/s*. Simultaneous profile and plan views are presented but each row is, in fact, a different bubble. The flow is from right to left. The scale is 4.5 times lifesize. From Ceccio and Brennen (1991).

Two other processes are occuring at the same time. First, the streamwise thickness of the bubble decreases faster than its spanwise breadth (spanwise being defined as the direction parallel to the headform surface and normal to the oncoming stream), so that the largest dimension of the bubble is its spanwise breadth. Second, the bubble acquires significant spanwise vorticity through its interactions with the boundary layer during the growth phase. Consequently, as the collapse proceeds, this vorticity is concentrated and the bubble evolves into one (or two or possibly more) cavitating vortex with a spanwise axis. These vortex bubbles proceed to collapse and seem to rebound as a cloud of much smaller bubbles. Often

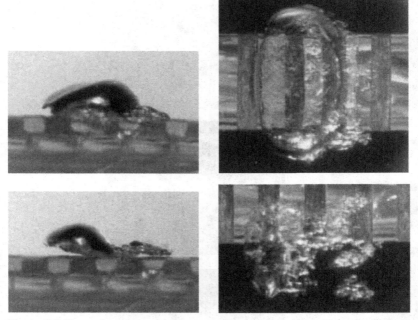

Figure 7.3 Examples of simultaneous profile and plan views illustrating the instability of the liquid layer under a traveling cavitation bubble. From Ceccio and Brennen (1991) experiments with a 5.08 *cm* diameter ITTC headform at $\sigma = 0.45$ and a speed of 8.7 *m/s*. The flow is from right to left and the scale is 3.8 times lifesize.

a coherent second collapse of this cloud was observed when the bubbles were not too scattered by the flow. Ceccio and Brennen (1991) (see also Kumar and Brennen 1993) conclude that the flow-induced fission prior to collapse can have a substantial effect on the noise impulse (see Section 3.8).

Two additional phenomena were observed on the ITTC headform, which exhibited laminar separation. The first of these was the observation that the layer of liquid underneath the bubble would become disrupted by some instability. As seen in Figure 7.3, this results in a bubbly layer of fluid that subsequently gets left behind the main bubble. Thus the instability of the liquid layer leads to another process of bubble fission. Because of the physical separation, the bubbly layer would collapse after the main body of the bubble.

The second and perhaps more consequential phenomenon observed with the ITTC headform only occurs with the occasional bubble. Infrequently, when a bubble passes the point of laminar separation, it triggers the formation of local "attached cavitation" streaks at the lateral or spanwise extremities of the bubble, as seen in Figure 7.4. Then, as the main bubble proceeds downstream, these "streaks" or "tails" of attached cavitation are stretched out behind the main bubble, the trailing ends of the tails being attached to the solid surface. Subsequently, the main bubble collapses first, leaving the "tails" to persist for a fraction longer, as illustrated by the lower photograph in Figure 7.4.

Figure 7.4 Examples illustrating the attached tails formed behind a traveling cavitation bubble. The top two are simultaneous profile and plan views. The bottom shows the persistence of the tails after the bubble has collapsed. From Ceccio and Brennen (1991) experiments with a 5.08 *cm* diameter ITTC headform at $\sigma = 0.42$ and a speed of 9 *m/s*. The flow is from right to left and the scale is 3.8 times lifesize.

The importance of these occasional "events with tails" did not become clear until tests were conducted at much higher Reynolds numbers, with larger headforms (up to 50.5 *cm* in diameter) and somewhat higher speeds (up to 15 *m/s*). These tests were part of an investigation of the scaling of the bubble dynamic phenomena described above (Kuhn de Chizelle et al. 1992a,b). One notable observation was the presence of a "dimple" on the exterior surface of all the individual traveling bubbles; examples of this dimple are included in Figure 7.5. They are not the precursor to a reentrant jet, for the dimple seems to be relatively stable during most of the collapse process. More importantly, it was observed that, at higher Reynolds number, "attached tails" occurred even on these Schiebe bodies, which did not normally exhibit laminar separation. Moreover, the probability of occurence of attached tails increased as the Reynolds number increased and the attached cavitation began to be more extensive. As the Reynolds number increased further, the bubbles would tend to trigger attached cavities over the entire wake of the bubble as seen in the lower two photographs in Figure 7.5. Moreover, the attached cavitation would tend to remain for a longer period after the main bubble had disappeared. Eventually, at the highest Reynolds numbers tested, it appeared that the passage of a single bubble was sufficient to trigger a "patch" of attached cavitation

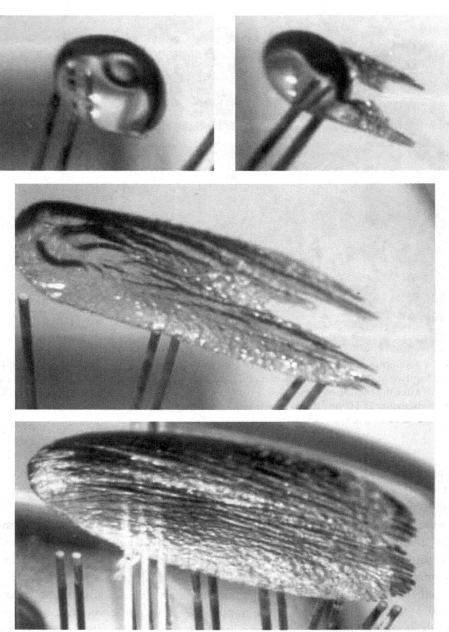

Figure 7.5 Typical cavitation events from the scaling experiments of Kuhn de Chizelle et al. (1992b) showing an unattached bubble with "dimple"(upper left), a bubble with attached tails (upper right), and a transient bubble-induced patch (middle), all occurring on the 50.8 *cm* diameter Schiebe headform at $\sigma = 0.605$ and a speed of 15 *m/s*. The bottom photograph shows a patch on the 25.4 *cm* headform at $\sigma = 0.53$ and a speed of 15 *m/s*. The flow is from right to left. The top three are shown at 1.3 times lifesize and the bottom at 1.25 times lifesize.

(Figure 7.5, bottom), which would persist for an extended period after the bubble had long disappeared. This progression of events and the changes in the probabilities of the different kinds of events with Reynolds number imply a rich complexity in the micro-fluidmechanics of cavitation bubbles, much of which remains to be understood. Its importance lies in the fact that these different types of events cause differences in the collapse process which, in turn, alters the noise produced (see Kuhn de Chizelle et al. 1992b) and, in all probability, the potential for cavitation damage. For example, the events with attached tails were found to produce significantly less noise than the events without tails. Due to the changes in the probabilities of occurence of these events with Reynolds number, this implies a scaling effect that had not been previously recognized. It also suggests some possible strategies for the reduction of cavitation noise and damage.

When examined in retrospect, one can identify many of these phenomena in earlier photographic observations, including the pioneering, high-speed movies taken by Knapp. As previously noted, Knapp and Hollander (1948), Parkin (1952), and others noted the spherical-cap shape of most traveling cavitation bubbles. The ITTC experiments (Lindgren and Johnsson 1966) emphasized the diversity in the kinds of cavitation events that could occur on a given body, and later authors attempted to identify, understand, and classify this spectrum of events. For example, Holl and Carroll (1979) observed a variety of different types of cavitation events on axisymmetric bodies and remarked that both traveling and attached cavitation "patches" occurred and could be distinguished from traveling bubble cavitation. A similar study of the different types of cavitation events was reported by Huang (1979), whose "spots" are synonymous with "patches."

7.5 Large-Scale Cavitation Structures

When the density of cavitation events becomes large enough, they begin to interact and to alter the flow in a significant way. This increase in density may come about as a result of a decrease in the cavitation number, which causes the activation of increasingly smaller nuclei, or it may result from an increase in the population of nuclei in the oncoming stream. As long as the interaction effects are small, they seem to cause a decrease in the rate of growth of the bubbles (see, for example, Arakeri and Shanmuganathan 1985) and a shift in the spectrum of the cavitation noise (see, for example, Marboe, Billet, and Thompson 1986). Significant progress has been made in developing analytical models that incorporate such weak interaction effects on traveling bubble cavitation; these models are described in chapter 6.

An example of dense traveling bubble cavitation is included in Figure 7.6. Note that the bubbles seem to merge to form a single vapor-filled wake near the trailing edge of the foil. Notice also the wispy trails of very small air bubbles that remain after the vapor-filled cavity collapses. In a water tunnel special efforts are required to allow these fine bubbles sufficient time to dissolve before they recirculate back to the working section. Without such efforts the population of small bubbles in the tunnel would quickly reach unacceptable levels. Even with special efforts it is clear

Figure 7.6 Dense traveling bubble cavitation on the surface of a NACA 4412 hydrofoil at zero incidence angle, a speed of 13.7 *m/s* and a cavitation number of 0.3. The flow is from left to right and the leading edge of the foil is just to the left of the white glare patch on the surface (Kermeen 1956).

that cavitation itself contributes to the population of nuclei in a closed loop water tunnel.

The large-scale cavitation structures that are formed when the cavitation number is reduced can take a variety of forms, and we review these in the next few sections. In many practical devices such as pumps or propellers, the first large-scale structure to be observed as the cavitation number is decreased takes the form of a cavitating vortex, so we begin with a discussion of vortex cavitation.

7.6 Vortex Cavitation

Many high Reynolds number flows of practical importance contain a region of concentrated vorticity where the pressure in the vortex core is often significantly smaller than in the rest of the flow. Such is the case, for example, in the tip vortices of ship's propellers or pump impellers or in the swirling flow in the draft tube of a water turbine. It follows that cavitation inception often occurs in these vortices and that, with further reduction of the cavitation number, the entire core of the vortex may become filled with vapor. Naturally, the term "vortex cavitation" is used for these circumstances. In Figures 7.7 to 7.12 we present some examples of this particular kind of large-scale cavitation structure. Figure 7.7 consists of photographs of cavitating tip vortices on a finite aspect ratio hydrofoil at an angle of attack. In those experiments of Higuchi, Rogers, and Arndt (1986) cavitation inception occurred in the vortex some distance downstream of the tip at a cavitation number of about $\sigma = 1.4$. With further decrease in pressure the cavitation in the core becomes continuous, as illustrated by the picture on the left in Figure 7.7. This transition is probably triggered by an accumulation of individual bubbles in the core; they will tend to

Figure 7.7 Cavitating tip vortices generated by a finite aspect ratio hydrofoil of ellipsoidal planform at an angle of attack. On the left is a continuous tip vortex cavity at a cavitation number, $\sigma = 1.15$, and an angle of attack of $7.5°$. On the right, the tip vortex emerges from some surface cavitation at a lower value of $\sigma = 0.43$ (angle of attack = $9.5°$). Reproduced from Higuchi, Rogers, and Arndt (1986) with the authors' permission.

Figure 7.8 Cavitating tip vortex on a scale model of the low-pressure LOX turbopump impeller in the Space Shuttle Main Engine. The fluid is water, the inlet flow coefficient is 0.07 and the cavitation number is 0.42. Reproduced from Braisted (1979).

migrate to the center of the vortex due to the centrifugal pressure gradient. With further decrease in σ, bubble and/or sheet cavitation appear on the hydrofoil surface (Figure 7.7, photograph on right) and disturb the tip vortex which is nevertheless still apparent. Cavitating tip vortices are also quite apparent in unshrouded pump impellers as illustrated by Figure 7.8.

When continuous cavitating tip vortices occur at the tips of the blades of a propeller they create a surprisingly stable flow structure. As illustrated by Figure 7.9 the intertwined, helical cavitating vortices from the blade tips can persist for a long distance downstream of the propeller.

Clearly cavitation can occur in any vortex, and Figures 7.10 and 7.11 present two further examples. Figure 7.10 shows a typical picture of a cavitating vortex in

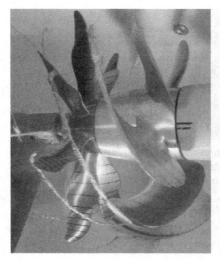

Figure 7.9 Tip vortex cavitation on a model propeller. Reproduced with permission of the Netherlands Maritime Research Institute and Lips B.V.

Figure 7.10 Cavitating vortex in the draft tube of a Francis turbine. Reproduced with the permission of P.Henry, Institut de Machines Hydrauliques et de Mecanique de Fluides, Ecole Polytechnique Federal de Lausanne, Switzerland.

the swirling flow in the draft tube of a Francis turbine. Often these draft tube vortices can exhibit quite complex patterns of unsteady flow. The vortices in a turbulent mixing layer or wake will also cavitate, as illustrated in Figure 7.11, a photograph of the separated wake behind a lifting flat plate with a flap. Looking closely at the structures in this turbulent flow, one can identify not only the large transverse

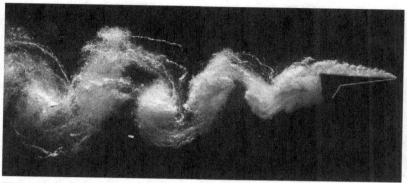

Figure 7.11 Cavitating vortices in the separated wake of a lifting flat plate with a flap; the flow is from the right to the left. Reproduced with the permission of A.J. Acosta.

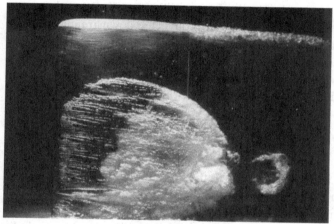

Figure 7.12 The formation of a ring vortex in the closure region of an attached cavity on an oscillating, finite-aspect-ratio hydrofoil with a chord of $0.152\ m$. The incidence angle is oscillating between $5°$ and $9°$ at a frequency of $10\ Hz$. The flow is from left to right at a velocity of $8.5\ m/s$ and a mean cavitation number of 0.5. Note the cavitating tip vortex as well as the attached cavity. Photograph by D.P. Hart.

vortices that contain many bubbles, but also the filament-like longitudinal vortices first identified in a single-phase mixing layer flow by Bernal and Roshko (1986). After that discovery by Bernal and Roshko one could recognize this secondary vortex structure in photographs of cavitating wakes and mixing layers taken many years previously, and yet its importance was not appreciated at the time. The streamwise vortices can play a particularly important role in cavitation inception. Katz and O'Hern (1986) have shown that, when streamwise vortices are present, inception occurs in these longitudinal structures before it occurs in the primary or transverse vortices.

The three-dimensional shedding of vortices from a finite aspect ratio foil or other device can often lead to the formation and propagation of a ring vortex with a vapor/gas core. Figure 7.12 shows such a cavitating vortex ring that has just emerged

Figure 7.13 A vortex ring shed by the partial cavitation oscillations of a hydrofoil. The flow is from right to left. Reproduced with the permission of A.J. Acosta.

from the closure region of an attached cavity on an oscillating foil. Often these ring vortices can persist for quite a distance as they are convected downstream. Another example is shown in Figure 7.13; in this case the vortex shedding is caused by the natural oscillations of a partially cavitating foil (see Section 7.9). The cavitating ring vortex has its own velocity of propagation relative to the surrounding fluid and has therefore moved substantially above the rest of the wake at the moment when the photograph was taken.

7.7 Cloud Cavitation

In many flows of practical interest one observes the periodic formation and collapse of a "cloud" of cavitation bubbles. Such a structure is termed "cloud cavitation." The temporal periodicity may occur naturally as a result of the shedding of cavitating vortices (see, for example, Figure 7.11), or it may be the response to a periodic disturbance imposed on the flow. Common examples of imposed fluctuations are the interaction between rotor and stator blades in a pump or turbine and the interaction between a ship's propeller and the nonuniform wake created by the hull. In many of these cases the coherent collapse of the cloud of bubbles (see, for example, Figure 3.14) can cause more intense noise and more potential for damage than in a similar nonfluctuating flow (see Section 3.7). Bark and van Berlekom (1978), Shen and Peterson (1978), Franc and Michel (1988), Kubota et al. (1989), and Hart et al. (1990) have studied the complicated flow patterns involved in the production and collapse of a cavitating cloud on an oscillating hydrofoil. These studies are exemplified by the photographs of Figure 7.14, which show the formation, separation, and collapse of a cavitation cloud on a hydrofoil oscillating in pitch. All of these studies emphasize that a substantial bang occurs as a result of the collapse of the cloud; in Figure 7.14 this occurred between the middle and right-hand photographs.

Cloud cavitation continues to be a primary concern for propeller and pump manufacturers and is currently the subject of active research. In Chapter 6 we presented some simplified, analytical investigations that provided some qualitative

Figure 7.14 Three frames illustrating the formation, separation, and collapse of a cavitation cloud on the suction surface of a hydrofoil (0.152 *m* chord) oscillating in pitch with a frequency of 5.8 *Hz* and an amplitude of ±5° about a mean incidence angle of 5°. The flow is from left to right, the tunnel velocity is 7.5 *m/s* and the mean cavitation number is 1.1. Photographs by E.McKenney.

information on the coherent dynamics of these structures. More accurate modeling of these complex, unsteady multiphase flows poses some challenging problems that have only begun to be addressed. The recent numerical modeling by Kubota, Kato, and Yamaguchi (1992) is an important step in this direction.

7.8 Attached or Sheet Cavitation

Another class of large-scale cavitation structures is that which occurs when a wake or region of separated flow fills with vapor. Referring back to Figure 7.6, we note that Kermeen (1956) only observed dense traveling bubbles when the angle of attack was small. At angles of attack greater than about 10° (or less than about −2°) cavitation occurred as a single vapor-filled separation zone as illustrated in Figure 7.15. This form of cavitation on a hydrofoil or propeller blade is usually termed "sheet" cavitation; in the context of pumps it is known as "blade" cavitation.

Bluff bodies often exhibit a sudden transition from traveling bubble cavitation to a single vapor-filled wake as the cavitation number is decreased. An example is shown in Figure 7.16 which includes two photographs of a cavitating sphere; the transition occurs when the bubbly wake in the picture on the left suddenly becomes a single vapor-filled void as seen in the picture on the right. In the context of bluff bodies, a vapor-filled wake is often called a "fully developed" or "attached" cavity. Clearly sheet, blade, fully developed, and attached cavities are terms for the same large-scale cavitation structure.

When a sharp edge provides a clean definition for the leading edge of a fully developed cavity, the surface of that cavity is often glassy smooth since the separating boundary layer is usually laminar. This initially smooth surface can be seen in the right-hand photograph of Figure 7.16 and in the photographs of Figure 7.17. Depending on the shape of the forebody the interfacial boundary layer may rapidly undergo transition to a turbulent interfacial layer, as is the case in the photograph of the cavitating ogive in Figure 7.17 and the cavitating sphere in the photograph on the right of Figure 7.16. For other headforms transition may be delayed

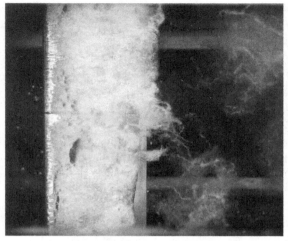

Figure 7.15 Sheet cavitation on the suction surface of a NACA 4412 hydrofoil at an angle of attack of 12°, a speed of 10.7 *m/s* and a cavitation number of 1.05 (Kermeen 1956). The flow is from left to right.

Figure 7.16 Two photographs of a cavitating, 7.62 *cm* diameter sphere. The left photographs shows bubble cavitation and bubbly wake prior to the transition to the fully developed cavity shown on the right (Brennen 1970). The flow is from right to left, the velocities being 5.6 *m/s* and 10.7 *m/s*, respectively.

Figure 7.17 Two fully developed cavities on a 5.95 *cm* diameter ogive (left) and a 7.62 *cm* diameter disc (right) set normal to the oncoming stream (Brennen 1970). The flow is from right to left and the velocities are 7.62 *m/s* and 10.7 *m/s*, respectively.

almost indefinitely, as in the case of the cavitating disc of Figure 7.17 (see Brennen 1970).

When there is no sharp edge to initiate a fully developed cavity, several different phenomena may occur. Cavitation separation may still occur along a well-defined and stable line on the body surface, as exemplified by the photograph on the right in Figure 7.16. Or the separation line may be interrupted, as in the photograph of Figure 7.18. For example, such a scalloped leading edge is typical of cavitation in bearings (Dowson and Taylor 1979).

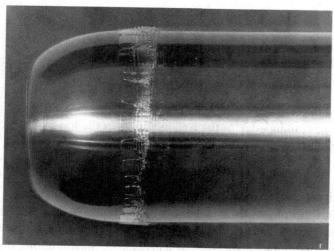

Figure 7.18 Sheet cavitation on the ITTC headform. The flow is from left to right with a speed of 12.2 *m/s* and a cavitation number of 0.424. Reproduced with the permission of A.J. Acosta.

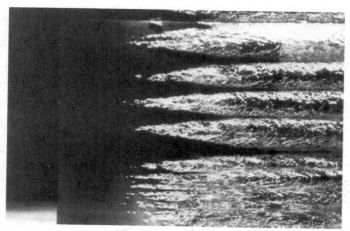

Figure 7.19 Streak cavitation on a biconvex hydrofoil at a speed of 15.5 *m/s* and a cavitation number of 0.11 (Arakeri 1975). The flow is from left to right and the leading edge of the foil is about 1 *cm* from the left-hand edge. Reproduced with the permission of V.H. Arakeri.

Other forms of developed cavitation can be strikingly different from that of Figures 7.17 or 7.18. Sometimes the cavities occur as streaks, as exemplified by the photograph in Figure 7.19 of cavitation on the surface of a biconvex hydrofoil (Arakeri 1975). Again a tranverse periodicity appears to occur in which one can envisage that the expansion of the flow in the streamtubes containing cavities results in an increase in the pressure in the fluid in between these cavitating stream-tubes and therefore inhibits further lateral spreading of the cavitation. Currently there does not appear to be any clear understanding of the reason for the transverse periodicity of Figures 7.18 and 7.19.

7.9 Cavitating Foils

On a lifting foil (a hydrofoil), attached cavitation can take a number of forms, as discussed in the review by Acosta (1973). When, as sketched in Figure 7.20, the attached cavity closes on the suction surface of the foil, the condition is referred to as "partial cavitation." This is the form of attached cavitation most commonly observed on propellers and in pumps. At lower cavitation numbers, the cavity may close well downstream of the trailing edge of the foil, as shown in the lower sketch in Figure 7.20. Such a configuration is termed "super-cavitation" and propellers for high-speed boats are often designed to be operated under these conditions. In between these regimes, experiments have shown (Wade and Acosta 1966) that, when the length of the cavity is close to the length of the foil (between about 3/4 and 4/3 times the chord), the flow becomes unstable and the size of the cavity fluctuates quite violently between these limits. During this fluctuation cycle, the cavity lengthens fairly smoothly. On the other hand, it shortens by a process of "pinching-off" of a large cloud of bubbles from the rear of the cavity, and this cloud can collapse quite violently as described previously. However, there is also shed vorticity bound up in the cloud, and this is concentrated by the collapse of the cloud. One result is the

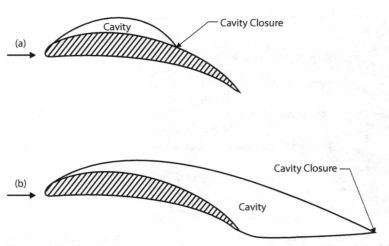

Figure 7.20 Sketch of the types of attached cavitation on a lifting foil: (a) partial cavitation (b) supercavitation.

formation of the vortex ring seen in Figure 7.13. In pumps and other devices, this condition between partial and supercavitation clearly needs to be avoided because of the potential damage that can result. Further discussion of this oscillating cavity phenomenon is included in Section 8.8. It should also be noted that cavities may fluctuate for other reasons, as discussed in the next section.

Methods for the analysis of both partially and supercavitating flows are discussed in the next chapter.

7.10 Cavity Closure

The flow in the vicinity of cavity closure deserves further comment because it is quite complex and involves processes that have not, as yet, been discussed. First, the flow is invariably turbulent since the boundary layer, which detaches from the body along with the free surface, produces an interfacial boundary layer. This is almost always unstable and undergoes transition to yield a turbulent interfacial layer (Brennen 1970). The level of turbulence in this layer grows rapidly as the closure region is approached, so the flow in that vicinity usually appears as a frothy turbulent mixing motion. Where the two free surface streams collide, some flow is deflected back into the cavity. Observations of this "reentrant jet" were part of the motivation for the reentrant jet model of cavity closure, which is sometimes employed in potential flow solutions (see Section 8.2). However, actual reentrant jets are nothing like as coherent as the jet in that model; they could better be described as a frothy turbulent mass tumbling back into the cavity.

Changes to the structure of the flow in the closure region can occur in horizontal flows when the buoyancy forces become significant. Such will be the case when the Froude number based on cavity length, ℓ, $Fr = U_\infty/(g\ell)^{\frac{1}{2}}$, is less than some critical value denoted by Fr_c. For bodies of small aspect ratio (such as axisymmetric headforms) it appears that $Fr_c \approx 2.5$ (Brennen 1969) and, when $Fr < Fr_c$, the reentrant jet structure no longer occurs. Instead, a pair of counter-rotating vortices with gas/vapor cores form in the closure region (Cox and Claydon 1956); this type of closure is much steadier and less turbulent than the reentrant jet type, which is prevalent at higher Froude numbers. The rate at which vapor/gas can be entrained by the counter-rotating vortex closure is much higher than for the reentrant jet closure (Brennen 1969).

Returning to our discussion of the reentrant jet form of cavity closure, we note that this flow can also exhibit significant fluctuations. These fluctuations can be caused by vortex shedding from the rear of the cavity (Young and Holl 1966); they may also be the result of some other, less well understood instability associated with this complex multiphase flow. Knapp (1955) first described the cyclic process in which a "pinching off" mechanism (similar to that described in the last section) produces vortices that initially have large, bubbly vapor/gas cores (see also Furness and Hutton 1975). As the vapor condenses and the core of the cloud/vortex collapses, the vorticity is concentrated and the vortices become more intense before they enter the normal, single-phase wake flow. After condensation, only small, remnant gas bubbles containing the residual noncondensable component remain to be convected away into the far wake. It is, incidentally, this supply of microbubbles to the tunnel population that neccessitates the use of a resorber in a cavitation tunnel (see Section 1.15).

It should also be noted that under some circumstances this cyclic process in the cavity closure region is more evident than in others. Moreover, there are several other instabilities that can trigger or promote such a cyclic shedding process. We have already discussed one such instability in the preceding section, the partial cavitation instability. A somewhat similar cavity pulsation phenomenon occurs when large super-cavities are created by supplying noncondensable gas to the wake of a body. Such cavities, which are visually almost indistinguishable from their natural or vapor-filled counterparts, are known as "ventilated" cavities. However, when the gas supplied is increased to the point at which the entrainment processes in the closure region (see below) are unable to carry away that volume of gas, the cavity may begin to fluctuate; a pinching-off process sheds a large gas volume into the wake, and this is followed by regrowth of the cavity. This phenomenon was investigated by Silberman and Song (1961) and Song (1962). Finally, we should mention one other process that may be at work in the closure region. In the case of predominantly vapor-filled cavities Jakobsen (1964) has suggested that a condensation shock provides a mechanism for cavity closure (simple shocks of this kind were analysed in Section 6.9). This last suggestion deserves more study than it has received to date.

Both the large-scale fluctuations and the small-scale turbulence in the closure region act to entrain bubbles and thus remove vapor/gas from the cavity, though it is clear from the preceding paragraphs that the precise mechanisms of entrainment may differ considerably from one closure configuration to another. Measurements of the volume rate of entrainment for large cavities with the steady, reentrant jet type of closure (for example, Brennen 1969) suggest that the volume rate increases with velocity as U_∞^n where n is a little larger than unity. Using axisymmetric headforms of different size, b, Billet and Weir (1975) showed that though the volume entrainment rate scaled approximately with $U_\infty b^2$, there was a significant variation with cavitation number, σ, the volume rate increasing substantially as σ decreased and the cavity became larger.

Under steady-state conditions, the removal of vapor and noncondensable gas by entrainment in the closure region is balanced by the supply process of evaporation and the release of gas from solution along the length of the free surface. These supply processes will, in turn, be affected by the state of the interfacial boundary layer. A turbulent layer will clearly enhance the heat and mass diffusion processes that produce evaporation and the release of gas from solution. One of the consequences of the balance between the supply of noncondensable gas (air) and its removal by entrainment is the inherent regulation of the partial pressure of the noncondensable gas (air) in the cavity. Brennen (1969) put together a simplified model of these processes and showed that the results for the partial pressure of air were in rough agreement with experimental measurements of that partial pressure.

Moreover, there is an analogous balance of heat in which the latent heat removed by the entrainment process must be balanced by the heat diffused to the cavity through the interfacial layer. This requires a cavity temperature below that of the surrounding liquid. (This thermal effect in fully developed cavity flows is analogous to the thermal effect in the dynamics of individual bubbles described in Sections 2.3 and 2.7.) The temperature depression produced by this process has been investigated by a number of authors including Holl, Billet, and Weir (1975). Though it is usually small in water at normal temperatures, it can be significant at higher temperatures or in other liquids at temperatures similar to those at which single bubbles experience significant thermal effects on growth (see Section 2.7).

References

Acosta, A.J. (1973). Hydrofoils and hydrofoil craft. *Ann. Rev. Fluid Mech.*, **5**, 161–184.

Arakeri, V.H. and Acosta, A.J. (1973). Viscous effects in the inception of cavitation on axisymmetric bodies. *ASME J. Fluids Eng.*, **95**, 519–528.

Arakeri, V.H. (1975). Viscous effects on the position of cavitation separation from smooth bodies. *J. Fluid Mech.*, **68**, 779–799.

Arakeri, V.H. (1979). Cavitation inception. *Proc. Indian Acad. Sci.*, **C2**, Pt. 2, 149–177.

Arakeri, V.H. and Shanmuganathan, V. (1985). On the evidence for the effect of bubble interference on cavitation noise. *J. Fluid Mech.*, **159**, 131–150.

Bark, G. and van Berlekom, W.B. (1978). Experimental investigations of cavitation noise. *Proc. 12th ONR Symp. on Naval Hydrodynamics*, 470–493.

Bernal, L.P. and Roshko, A. (1986). Streamwise vortex structure in plane mixing layers. *J. Fluid Mech.*, **170**, 499–525.

Billet, M.L. and Weir, D.S. (1975). The effect of gas diffusion on the flow coefficient for a ventilated cavity. *Proc. ASME Symp. on Cavity Flows*, 95–100.

Birkhoff, G. and Zarantonello, E.H. (1957). *Jets, wakes and cavities.* Academic Press.

Blake, W. K., Wolpert, M. J., and Geib, F. E. (1977). Cavitation noise and inception as influenced by boundary-layer development on a hydrofoil. *J. Fluid Mech.*, **80**, 617–640.

Braisted, D.M. (1979). Cavitation induced instabilities associated with turbomachines. *Ph.D. Thesis, Calif. Inst. of Tech.*

Brennen, C. (1969). The dynamic balances of dissolved air and heat in natural cavity flows. *J. Fluid Mech.*, **37**, 115–127.

Brennen, C. (1970). Cavity surface wave patterns and general appearance. *J. Fluid Mech.*, **44**, 33–49.

Briançon-Marjollet, L., Franc, J.P., and Michel, J.M. (1990). Transient bubbles interacting with an attached cavity and the boundary layer. *J. Fluid Mech.*, **218**, 355–376.

Ceccio, S. L. and Brennen, C. E. (1991). Observations of the dynamics and acoustics of travelling bubble cavitation. *J. Fluid Mech.*, **233**, 633–660.

Cox, R.N. and Claydon, W.A. (1956). Air entrainment at the rear of a steady cavity. *Proc. Symp. on Cavitation in Hydrodynamics, N.P.L., London.*

Dowson, D. and Taylor, C.M. (1979). Cavitation in bearings. *Ann. Rev. of Fluid Mech.*, **11**, 35–66.

Ellis, A.T. (1952). Observations on bubble collapse. *Calif. Inst. of Tech. Hydro. Lab. Rep. 21–12.*

Franc, J.P. and Michel, J.M. (1988). Unsteady attached cavitation on an oscillating hydrofoil. *J. Fluid Mech.*, **193**, 171–189.

Furness, R.A. and Hutton, S.P. (1975). Experimental and theoretical studies on two-dimensional fixed-type cavities. *Proc. ASME Symp. on Cavity Flows*, 111–117.

Hart, D.P., Brennen, C.E., and Acosta, A.J. (1990). Observations of cavitation on a three-dimensional oscillating hydrofoil. *ASME Cavitation and Multiphase Flow Forum*, **FED 98**, 49–52.

Higuchi, H., Rogers, M.F., and Arndt, R.E.A. (1986). Characteristics of tip cavitation noise. *Proc. ASME Int. Symp. on Cavitation and Multiphase Flow Noise*, **FED 45**, 101–106.

Holl, J.W. (1969). Limited cavitation. In *Cavitation State of Knowledge* (eds: J.M. Robertson and G.F. Wislicenus), ASME, New York.

Holl, J.W. and Kornhauser, A.L. (1970). Thermodynamic effects on desinent cavitation on hemispherical nosed bodies in water at temperatures from $80°F$ to $260°F$. *ASME J. Basic Eng.*, **92**, 44–58.

Holl, J.W., Billet, M.L., and Weir, D.S. (1975). Thermodynamic effects on developed cavitation. *Proc. ASME Symp. on Cavity Flows*, 101–109.

Holl, J.W. and Carroll, J.A. (1979). Observations of the various types of limited cavitation on axisymmetric bodies. *Proc. ASME Int. Symp. on Cavitation Inception*, 87–99.

Huang, T.T. (1979). Cavitation inception observations on six axisymmetric headforms. *Proc. ASME Int. Symp. on Cavitation Inception*, 51–61.

Jakobsen, J.K. (1964). On the mechanism of head breakdown in cavitating inducers. *ASME J. Basic Eng.*, **86**, 291–304.

Johnson, V.E., Jr. and Hsieh, T. (1966). The influence of the trajectories of gas nuclei on cavitation inception. *Proc. 6th ONR Symp. on Naval Hydrodynamics*, 163–182.

Katz, J. and O'Hern, T.J. (1986). Cavitation in large scale shear flows. *ASME J. Fluids Eng.*, **108**, 373–376.

Kermeen, R.W. (1956). Water tunnel tests of NACA 4412 and Walchner Profile 7 hydrofoils in non-cavitating and cavitating flows. *Calif. Inst. of Tech. Hydro. Lab. Rep. 47–5.*

Kermeen, R.W., McGraw, J.T., and Parkin, B.R. (1955). Mechanism of cavitation inception and the related scale-effects problem. *Trans. ASME*, **77**, 533–541.

Kirchhoff, G. (1869). Zur Theorie freier Flüssigkeitsstrahlen. *Z. reine Angew. Math.*, **70**, 289–298.

Knapp, R.T. and Hollander, A. (1948). Laboratory investigations of the mechanism of cavitation. *Trans. ASME*, **70**, 419–435.

Knapp, R.T. (1955). Recent investigations of the mechanics of cavitation and cavitation damage. *Trans. ASME*, **77**, 1045–1054.

Kubota, A., Kato, H., Yamaguchi, H., and Maeda, M. (1989). Unsteady structure measurement of cloud cavitation on a foil section using conditional sampling. *ASME J. Fluids Eng.*, **111**, 204–210.

Kubota, A., Kato, H., and Yamaguchi, H. (1992). A new modelling of cavitating flows—a numerical study of unsteady cavitation on a hydrofoil section. *J. Fluid Mech.*, **240**, 59–96.

Kuhn de Chizelle, Y., Ceccio, S.L., Brennen, C.E., and Gowing, S. (1992). Scaling experiments on the dynamics and acoustics of travelling bubble cavitation. *Proc. 3rd I. Mech. E. Int. Conf. on Cavitation, Cambridge, England*, 165–170.

Kuhn de Chizelle, Y., Ceccio, S.L., Brennen, C.E., and Shen, Y. (1992). Cavitation scaling experiments with headforms: bubble acoustics. *Proc. 19th ONR Symp. on Naval Hydrodynamics*, 72–84.

Kumar, S. and Brennen, C.E. (1993). A study of pressure pulses generated by travelling bubble cavitation. *J. Fluid Mech.*, **255**, 541–564.

Lindgren, H. and Johnsson, C.A. (1966). Cavitation inception on headforms, ITTC comparative experiments. *Proc. 11th Int. Towing Tank Conf., Tokyo*, 219–232.

Marboe, R.C., Billet, M.L., and Thompson, D.E. (1986). Some aspects of travelling bubble cavitation and noise. *Proc. ASME Int. Symp. on Cavitation and Multiphase Flow Noise*, **FED 45**, 119–126.

Oshima, R. (1961). Theory of scale effects on cavitation inception on axially symmetric bodies. *ASME J. Basic Eng.*, **83**, 379–398.

Parkin, B.R. (1952). Scale effects in cavitating flow. *Ph.D. Thesis, Calif. Inst. of Tech.*

Plesset, M.S. (1948). The dynamics of cavitation bubbles. *ASME J. Appl. Mech.*, **16**, 228–231.

Schiebe, F.R. (1972). Measurements of the cavitation susceptibility using standard bodies. *St. Anthony Falls Hydr. Lab., Univ. of Minnesota, Rep. No. 118.*

Shen, Y. and Peterson, F.B. (1978). Unsteady cavitation on an oscillating hydrofoil. *Proc. 12th ONR Symp. on Naval Hydrodynamics*, 362–384.

Silberman, E. and Song, C.S. (1961). Instability of ventilated cavities. *J. Ship Res.*, **5**, 13–33.

Song, C.S. (1962). Pulsation of ventilated cavities. *J. Ship Res.*, **5**, 8–20.

Titchmarsh, E.C. (1947). *The theory of functions.* Oxford Univ. Press.

van der Meulen, J.H.J. and van Renesse, R.L. (1989). The collapse of bubbles in a flow near a boundary. *Proc. 17th ONR Symp. on Naval Hydrodynamics*, 379–392.

Van der Walle, F. (1962). On the growth of nuclei and the related scaling factors in cavitation inception. *Proc. 4th ONR Symp. on Naval Hydrodynamics*, 357–404.

Van Tuyl, A. (1950). On the axially symmetric flow around a new family of half-bodies. *Quart. Appl. Math.*, **7**, 399–409.

Wade, R.B. and Acosta, A.J. (1966). Experimental observations on the flow past a plano-convex hydrofoil. *ASME J. Basic Eng.*, **88**, 273–283.

Weinstein, A. (1948). On axially symmetric flow. *Quart. Appl. Math.*, **5**, 429–444.

Young, J.O. and Holl, J.W. (1966). Effects of cavitation on periodic wakes behind symmetric wedges. *ASME J. Basic Eng.*, **88**, 163–176.

8 Free Streamline Flows

8.1 Introduction

In this chapter we briefly survey the extensive literature on fully developed cavity flows and the methods used for their solution. The terms "free streamline flow" or "free surface flow" are used for those situations that involve a "free" surface whose location is initially unknown and must be found as a part of the solution. In the context of some of the multiphase flow literature, they would be referred to as *separated flows*. In the introduction to Chapter 6 we described the two asymptotic states of a multiphase flow, *homogeneous* and *separated* flow. Chapter 6 described some of the homogeneous flow methods and their application to cavitating flows; this chapter presents the other approach. However, we shall not use the term *separated flow* in this context because of the obvious confusion with the accepted, fluid mechanical use of the term.

Fully developed cavity flows constitute one subset of free surface flows, and this survey is intended to provide information on some of the basic properties of these flows as well as the methods that have been used to generate analytical solutions of them. A number of excellent reviews of free streamline methods can be found in the literature, including those of Birkhoff and Zarantonello (1957), Parkin (1959), Gilbarg (1960), Woods (1961), Gurevich (1961), Sedov (1966), and Wu (1969, 1972). Here we shall follow the simple and elegant treatment of Wu (1969, 1972).

The subject of free streamline methods has an interesting history, for one can trace its origins to the work of Kirchhoff (1869), who first proposed the idea of a "wake" bounded by free streamlines as a model for the flow behind a finite, bluff body. He used the mathematical methods of Helmholtz (1868) to find the irrotational solution for a flat plate set normal to an oncoming stream. The pressure in the wake was assumed to be constant and equal to the upstream pressure. Under these conditions (the zero cavitation number solution described below) the wake extends infinitely far downstream of the body. The drag on the body is nonzero, and Kirchhoff proposed this as the solution to D'Alembert's paradox (see Section 5.2), thus generating much interest in these free streamline methods, which Levi-Civita (1907) later extended to bodies with curved surfaces. It is interesting to note that Kirchhoff's work appeared many years before Prandtl discovered boundary layers and the reason for the wake structure behind a body. However, Kirchhoff made no mention of the possible application of his methods to cavity flows; indeed, the existence of these flows does not seem to have been recognized until many years later.

In this review we focus on the application of free streamline methods to fully developed cavity flows; for a modern view of their application to wake flows the reader is referred to Wu (1969, 1972). It is important to take note of the fact that, because of its low density relative to that of the liquid, the nature of the vapor or gas in the fully developed cavity usually has little effect on the liquid flow. Thus the pressure gradients due to motion of the vapor/gas are normally negligible relative to the pressure gradients in the liquid, and consequently it is usually accurate to assume that the pressure, p_c, acting on the free surface is constant. Similarly, the shear stress that the vapor/gas imposes on the free surface is usually negligible. Moreover, other than the effect on p_c, it is of little consequence whether the cavity contains vapor or noncondensable gas, and the effect of p_c is readily accommodated in the context of free streamline flows by defining the cavitation number, σ, as

$$\sigma = \frac{p_\infty - p_c}{\frac{1}{2} \rho_L U_\infty^2} \tag{8.1}$$

where p_c has replaced the p_V of the previous definition and we may consider p_c to be due to any combination of vapor and gas. It follows that the same free streamline analysis is applicable whether the cavity is a true vapor cavity or whether the wake has been filled with noncondensable gas externally introduced into the "cavity." The formation of such gas-filled wakes is known as "ventilation." Ventilated cavities can occur either because of deliberate air injection into a wake or cavity, or they may occur in the ocean due to naturally occurring communication between, say, a propeller blade wake and the atmosphere above the ocean surface. For a survey of ventilation phenomena the reader is referred to Acosta (1973).

Most of the available free streamline methods assume inviscid, irrotational and incompressible flow, and comparisons with experimental data suggest, as we shall see, that these are reasonable approximations. Viscous effects in fully developed cavity flows are usually negligible so long as the free streamline detachment locations (see Figure 8.1) are fixed by the geometry of the body. The most significant discrepancies occur when detachment is not fixed but is located at some initially unknown point on a smooth surface (see Section 8.3). Then differences between the calculated and observed detachment locations can cause substantial discrepancies in the results.

Assuming incompressible and irrotational flow, the problems require solution of Laplace's equation for the velocity potential, $\phi(x_i, t)$,

$$\nabla^2 \phi = 0 \tag{8.2}$$

subject to the following boundary conditions:

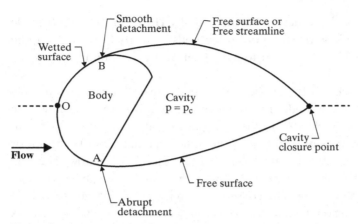

Figure 8.1 Schematic showing the terminology used in the free streamline analysis.

1. On a solid surface, $\mathcal{S}_W(x_i,t)$, the kinematic condition of no flow through that surface requires that

$$\frac{d\mathcal{S}_W}{dt} = \frac{\partial \mathcal{S}_W}{\partial t} + (\nabla\phi)\cdot\nabla\mathcal{S}_W = 0 \qquad (8.3)$$

2. On a free surface, $\mathcal{S}_F(x_i,t)$, a similar kinematic condition that neglects the liquid evaporation rate yields

$$\frac{d\mathcal{S}_F}{dt} = \frac{\partial \mathcal{S}_F}{\partial t} + (\nabla\phi)\cdot\nabla\mathcal{S}_F = 0 \qquad (8.4)$$

3. Assuming that the pressure in the cavity, p_c, is uniform and constant, leads to an additional dynamic boundary condition on $\mathcal{S}_F$. Clearly, the dimensionless equivalent of p_c, namely σ, is a basic parameter in this class of problem and must be specified *a priori*. In steady flow, neglecting surface tension and gravitational effects, the magnitude of the velocity on the free surface, q_c, should be uniform and equal to $U_\infty(1+\sigma)^{\frac{1}{2}}$.

The two conditions on the free surface create serious modeling problems both at the detachment points and in the cavity closure region (Figure 8.1). These issues will the addressed in the two sections that follow.

In planar, two-dimensional flows the powerful methods of complex variables and the properties of analytic functions (see, for example, Churchill 1948) can be used with great effect to obtain solutions to these irrotational flows (see the review articles and books mentioned above). Indeed, the vast majority of the published literature is devoted to such methods and, in particular, to steady, incompressible, planar potential flows. Under those circumstances the complex velocity potential, f, and the complex conjugate velocity, w, defined by

$$f = \phi + i\psi; \quad w = \frac{df}{dz} = u - iv \qquad (8.5)$$

are both analytic functions of the position vector $z = x + iy$ in the physical, (x,y) plane of the flow. In this context it is conventional to use i rather than j to denote $(-1)^{\frac{1}{2}}$ and we adopt this notation. It follows that the solution to a particular flow problem consists of determining the form of the function, $f(z)$ or $w(z)$. Often this takes a parametric form in which $f(\zeta)$ (or $w(\zeta)$) and $z(\zeta)$ are found as functions of some parametric variable, $\zeta = \xi + i\eta$. Another very useful device is the logarithmic hodograph variable, ϖ, defined by

$$\varpi = \log\frac{q_c}{w} = \chi + i\theta; \quad \chi = \ln\frac{q_c}{|w|}; \quad \theta = \tan^{-1}\frac{v}{u} \tag{8.6}$$

The value of this variable lies in the fact that its real part is known on a free surface, whereas its imaginary part is known on a solid surface.

8.2 Cavity Closure Models

Addressing first the closure problem, it is clear that most of the complex processes that occur in this region and that were described in Section 7.10 cannot be incorporated into a potential flow model. Moreover, it is also readily apparent that the condition of a prescribed free surface velocity would be violated at a rear stagnation point such as that depicted in Figure 8.1. It is therefore necessary to resort to some artifact in the vicinity of this rear stagnation point in order to effect termination of the cavity. A number of closure models have been devised; some of the most common are depicted in Figures 8.2 and 8.3. Each has its own advantages and deficiences:

1. Riabouchinsky (1920) suggested one of the simpler models, in which an "image" of the body is placed in the closure region so that the streamlines close smoothly onto this image. In the case of planar or axisymmetric bodies appropriate shapes for the image are readily found; such is not the case for general three-dimensional bodies. The advantage of the Riabouchinsky model is the simplicity of the geometry and of the mathematical solution. Since the combination of the body, its image, and the cavity effectively constitutes a finite body, it must satisfy D'Alembert's paradox, and therefore the drag force on the image must be equal and opposite to that on the body. Also note that the rear stagnation point is no longer located on a free surface but has been removed to the surface of the image. The deficiences of the Riabouchinsky model are the artificiality of the image body and the fact that the streamlines downstream are an image of those upstream. The model would be more realistic if the streamlines downstream of the body-cavity system were displaced outward relative to their locations upstream of the body in order to simulate the effect of a wake. Nevertheless, it remains one of the most useful models, especially when the cavity is large, since the pressure distribution and therefore the force on the body is not substantially affected by the presence of the distant image body.

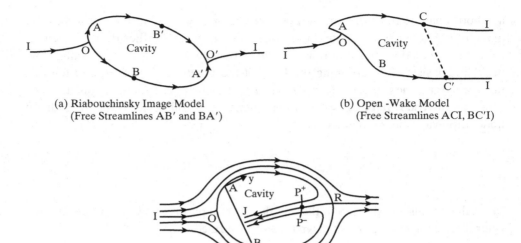

(a) Riabouchinsky Image Model
(Free Streamlines AB′ and BA′)

(b) Open -Wake Model
(Free Streamlines ACI, BC′I)

(c) Reentrant Jet Model
(Free Streamlines AJ, BJ)

Figure 8.2 Closure models for the potential flow around an arbitrary body shape (AOB) with a fully developed cavity having free streamlines or surfaces as shown. In planar flow, these geometries in the physical or z-plane transform to the geometries shown in Figure 8.11.

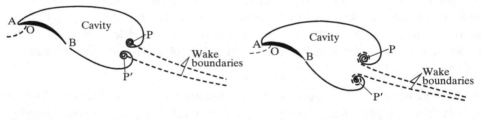

(d) Single Spiral Vortex Model

(e) Double Spiral Vortex Model

Figure 8.3 Two additional closure models for planar flow suggested by Tulin (1953, 1964). The free streamlines end in the center of the vortices at the points P and P' which are also the points of origin of the wake boundary streamlines on which the velocity is equal to U_∞.

2. Joukowski (1890) proposed solving the closure problem by satisfying the dynamic free surface condition only up to a certain point on the free streamlines (the points C and C' in Figure 8.2) and then somehow continuing these streamlines to downstream infinity, thus simulating a wake extending to infinity. This is known as the "open-wake model." For symmetric, pure-drag bodies these continuations are usually parallel with the uniform stream (Roshko 1954). Wu (1956, 1962) and Mimura (1958) extended this model to planar flows about lifting bodies for which the conditions on the continued streamlines are more complex. The advantage of the open-wake model is its simplicity. D'Alembert's paradox no longer applies since the effective body is now infinite. The disadvantage is that the wake is significantly larger than the real wake (Wu, Whitney, and

Brennen 1971). In this sense the Riabouchinsky and open-wake models bracket the real flow.

3. The "reentrant jet" model, which was first formulated by Kreisel (1946) and Efros (1946), is also shown in Figure 8.2. In this model, a jet flows into the cavity from the closure region. Thus the rear stagnation point, R, has been shifted off the free surfaces into the body of the fluid. Moreover, D'Alembert's paradox is again avoided because the effective body is no longer simple and finite; one can visualize the momentum flux associated with the reentrant jet as balancing the drag on the body. One of the motivations for the model is that reentrant jets are often observed in real cavity flows, as discussed in Section 7.10. In practice the jet impacts one of the cavity surfaces and is reentrained in an unsteady and unmodeled fashion. In the mathematical model the jet disappears onto a second Riemann sheet. This represents a deficiency in the model since it implies an unrealistic removal of fluid from the flow and consequently a wake of "negative thickness." In one of the few detailed comparisons with experimental observations, Wu et al. (1971) found that the reentrant jet model did not yield results for the drag that were as close to the experimental observations as the results for the Riabouchinsky and open-wake models.

4. Two additional models for planar, two-dimensional flow were suggested by Tulin (1953, 1964) and are depicted in Figure 8.3. In these models, termed the "single spiral vortex model" and the "double spiral vortex model," the free streamlines terminate in a vortex at the points P and P' from which emerge the bounding streamlines of the "wake" on which the velocity is assumed to be U_∞. The shapes of the two wake bounding streamlines are assumed to be identical, and their separation vanishes far downstream. The double spiral vortex model has proved particularly convenient mathematically (see, for example, Furuya 1975a) and has the attractive feature of incorporating a wake thickness that is finite but not as unrealistically large as that of the open-wake model. The single spiral vortex model has been extensively used by Tulin and others in the context of the linearized or small perturbation theory of cavity flows (see Section 8.7).

Not included in this list are a number of other closure models that have either proved mathematically difficult to implement or depart more radically from the observations of real cavities. For a discussion of these the reader is referred to Wu (1969, 1972) or Tulin (1964). Moreover, most of the models and much of the above discussion assume that the flow is steady. Additional considerations are necessary when modeling unsteady cavity flows (see Section 8.12).

8.3 Cavity Detachment Models

The other regions of the flow that require careful consideration are the points at which the free streamlines "detach" from the body. We use the word "detachment" to avoid confusion with the process of separation of the boundary layer. Thus the words "separation point" are reserved for boundary layer separation.

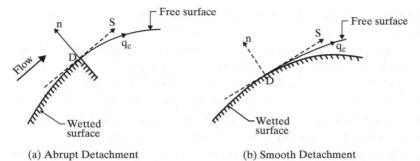

(a) Abrupt Detachment   (b) Smooth Detachment

Figure 8.4 Notation used in the discussion of the detachment of a free streamline from a solid body.

Since most of the mathematical models assume incompressible and irrotational potential flow, it is necessary to examine the prevailing conditions at a point at which a streamline in such a flow detaches from a solid surface. We first observe that if the pressure in the cavity is assumed to be lower than at any other point in the liquid, then the free surface must be convex viewed from the liquid. This precludes free streamlines with negative curvatures (the sign is taken to be positive for a convex surface). Second, we distinguish between the two geometric circumstances shown in Figure 8.4. Abrupt detachment is the term applied to the case in which the free surface leaves the solid body at a vertex or discontinuity in the slope of the body surface.

For convenience in the discussion we define a coordinate system, (s,n), whose origin is at the detachment point or vertex. The direction of the coordinate, s, coincides with the direction of the velocity vector at the detachment point and the coordinate, n, is perpendicular to the solid surface. It is sufficiently accurate for present purposes to consider the flow to be locally planar and to examine the nature of the potential flow solutions in the immediate neighborhood of the detachment point, D. Specifically, it is important to identify the singular behavior at D. This is most readily accomplished by using polar coordinates, (r,ϑ), where $z = s + in = r\,e^{i\vartheta}$, and by considering the expansion of the logarithmic hodograph variable, ϖ (Equation (8.6)), as a power series in z. Since, to first order, $Re\{\varpi\} = 0$ on $\vartheta = 0$ and $Im\{\varpi\} = 0$ on $\vartheta = \pi$, it follows that, in general, the first term in this expansion is

$$\varpi = -Ciz^{\frac{1}{2}} + \cdots \tag{8.7}$$

where the real constant C would be obtained as a part of the solution to the specific flow. From Equation (8.7), it follows that

$$w = q_c\{1 + Ciz^{\frac{1}{2}} + \cdots\} \tag{8.8}$$

$$f = \phi + i\psi = q_c\left\{z + \frac{2}{3}Ciz^{\frac{3}{2}} + \cdots\right\} \tag{8.9}$$

and the following properties of the flow at an abrupt detachment point then become evident. First, from Equation (8.8) it is clear that the acceleration of the fluid tends to infinity as one approaches the detachment point along the wetted surface. This, in turn, implies an infinite, favorable pressure gradient. Moreover, in order for the wetted surface velocity to be lower than that on the free surface (and therefore for the wetted surface pressure to be higher than that in the cavity), it is necessary for C to be a *positive* constant. Second, since the shape of the free surface, $\psi = 0$, is given by

$$y + Re\left\{\frac{2}{3}Cz^{\frac{3}{2}} + \cdots\right\} = 0 \qquad (8.10)$$

it follows that the curvature of that surface becomes infinite as the detachment point is approached along the free surface. The sign of C also implies that the free surface is convex viewed from within the liquid. The modifications to these characteristics as a result of a boundary layer in a real flow were studied by Ackerberg (1970); it seems that the net effect of the boundary layer on abrupt detachment is not very significant. We shall delay further discussion of the practical implications of these analytical results until later.

Turning attention to the other possibility sketched in Figure 8.4, "smooth detachment," one must first ask why it should be any different from abrupt detachment. The reason is apparent from one of the results of the preceding paragraph. An infinite, convex free-surface curvature at the detachment point is geometrically impossible at a smooth detachment point because the free surface would then cut into the solid surface. However, the position of the smooth detachment point is initially unknown. One can therefore consider a whole family of solutions to the particular flow, each with a different detachment point. There may be one such solution for which the strength of the singularity, C, is identically zero, and this solution, unlike all the others, is viable since its free surface does not cut into the solid surface. Thus the condition that the strength of the singularity, C, be zero determines the location of the smooth detachment point. These circumstances and this condition were first recognized independently by Brillouin (1911) and by Villat (1914), and the condition has become known as the Brillouin-Villat condition. Though normally applied in planar flow problems, it has also been used by Armstrong (1953), Armstrong and Tadman (1954), and Brennen (1969a) in axisymmetric flows.

The singular behavior at a smooth detachment point can be examined in a manner similar to the above analysis of an abrupt detachment point. Since the one-half power in the power law expansion of ϖ is now excluded, it follows from the conditions on the free and wetted surfaces that

$$\varpi = -Ciz^{\frac{3}{2}} + \cdots \qquad (8.11)$$

where C is a different real constant, the strength of the three-half power singularity. By parallel evaluation of w and f one can determine the following properties of the flow at a smooth detachment point. The velocity and pressure gradients approach zero (rather than infinity) as the detachment point is approached along the wetted surface. Also, the curvature of the free surface approaches that of the solid surface as the detachment point is approached along the free surface. Thus the name "smooth detachment" seems appropriate.

Having established these models for the detachment of the free streamlines in potential flow, it is important to emphasize that they are models and that viscous boundary-layer and surface-energy effects (surface tension and contact angle) that are omitted from the above discussions will, in reality, have a substantial influence in determining the location of the actual detachment points. This can be illustrated by comparing the locations of smooth detachment from a cavitating sphere with experimentally measured locations. As can readily be seen from Figures 8.5 and 8.6, the predicted detachment locations are substantially upstream of the actual detachment points. Moreover, the experimental data exhibit some systematic variations with the size of the sphere and the tunnel velocity. Exploring these scaling effects, Brennen (1969b) interpolated between the data to construct the variations with Reynolds number shown in Figure 8.6. This data clearly indicates that the detachment locations are determined primarily by viscous, boundary-layer effects. However, one must add that all of the experimental data used for Figure 8.6 was for metal spheres and that surface-energy effects and, in particular, contact-angle effects probably also play an important role (see Ackerberg 1975). The effect of the surface tension of the liquid seems to be relatively minor (Brennen 1970).

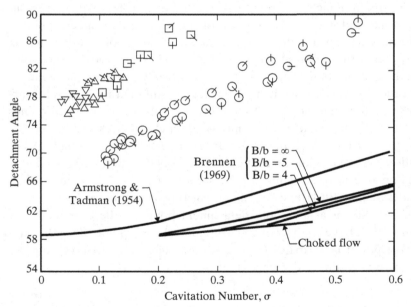

Figure 8.5 Observed and calculated locations of free surface detachment for a cavitating sphere. The detachment angle is measured from the front stagnation point. The analytical results using the smooth detachment condition are from Armstrong and Tadman (1954) and Brennen (1969a), in the latter case for different water tunnel to sphere radius ratios, B/b (see Figure 8.15). The experimental results are for different sphere diameters as follows: 7.62 cm ($\odot$) and 2.86 cm ($\square$) from Brennen (1969a), 5.08 cm ($\triangle$) and 3.81 cm ($\triangledown$) from Hsu and Perry (1954). Tunnel velocities are indicated by the additional ticks at cardinal points as follows: 4.9 m/s (NW), 6.1 m/s (N), 7.6 m/s (NE), 9.1 m/s (E), 10.7 m/s (SE), 12.2 m/s (S) and 13.7 m/s (SW).

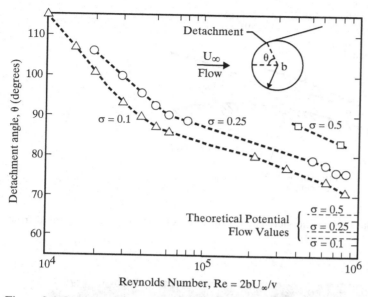

Figure 8.6 Observed free surface detachment points from spheres for various cavitation numbers, σ, and Reynolds numbers. Also shown are the potential flow values using the smooth detachment condition. Adapted from Brennen (1969b).

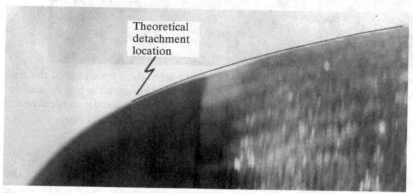

Figure 8.7 Comparison of the theoretical and experimental profiles of a fully developed cavity behind a sphere. The flow is from the right to the left. From Brennen (1969a).

It is worth noting that, despite the discrepancies between the observed locations of detachment and those predicted by the smooth detachment condition, the profile of the cavity is not as radically affected as one might imagine. Figure 8.7, taken from Brennen (1969a), is a photograph showing the profile of a fully developed cavity on a sphere. On it is superimposed the profile of the theoretical solution. Note the close proximity of the profiles despite the substantial discrepancy in the detachment points.

The viscous flow in the vicinity of an actual smooth detachment point is complex and still remains to be completely understood. Arakeri (1975) examined this

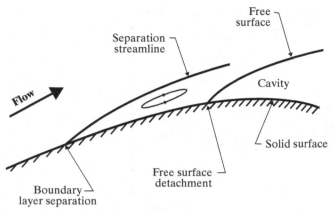

Figure 8.8 Model of the flow in the vicinity of a smooth detachment point. Adapted from Arakeri (1975).

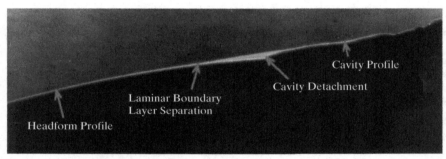

Figure 8.9 Schlieren photograph showing boundary layer separation upstream of the free surface detachment on an axisymmetric headform. The cavitation number is 0.39 and the tunnel velocity is $8.1\ m/s$. The actual distance between the separation and detachment points is about $0.28\ cm$. Reproduced from Arakeri (1975) with permission of the author.

issue experimentally using Schlieren photography to determine the behavior of the boundary layer and observed that boundary layer separation occurred upstream of free surface detachment as sketched in Figure 8.8 and shown in Figure 8.9. Arakeri also generated a quasi-empirical approach to the prediction of the distance between the separation and detachment locations, and this model seemed to produce detachment positions that were in good agreement with the observations. Franc and Michel (1985) studied this same issue both analytically and through experiments on hydrofoils, and their criterion for the detachment location has been used by several subsequent investigators.

In practice many of the methods used to solve free streamline problems involving detachment from a smooth surface simply assume a known location of detachment based on experimental observations (for example, Furuya and Acosta 1973) and neglect the difficulties associated with the resulting abrupt detachment solution.

8.4 Wall Effects and Choked Flows

Several useful results follow from the application of basic fluid mechanical principles to cavity flows constrained by uniform containing walls. Such would be the case, for example, for experiments in water tunnels. Consequently, in this section, we focus attention on the issue of wall effects in cavity flows and on the related phenomenon of *choked flow*. Anticipating some of the results of Figures 8.16 and 8.17, we observe that, for the same cavitation number, the narrower the tunnel relative to the body, the broader and longer the cavity becomes and the lower the drag coefficient. For a finite tunnel width, there is a critical cavitation number, σ_c, at which the cavity becomes infinitely long and no solutions exist for $\sigma < \sigma_c$. The flow is said to be *choked* at this limiting condition because, for a fixed tunnel pressure and a fixed cavity pressure, a minimum cavitation number implies an upper limit to the tunnel velocity. Consequently the choking phenomenon is analogous to that which occurs in a the nozzle flow of a compressible fluid (see Section 6.5). The phenomenon is familiar to those who have conducted experiments on fully developed cavity flows in water tunnels. When one tries to exceed the maximum, choked velocity, the water tunnel pressure rises so that the cavitation number remains at or above the choked value.

In the choked flow limit of an infinitely long cavity, application of the equations of conservation of mass, momentum, and energy lead to some simple relationships for the parameters of the flow. Referring to Figure 8.10, consider a body with a frontal projected area of A_B in a water tunnel of cross-sectional area, A_T. In the limit of an infinitely long cavity, the flow far downstream will be that of a uniform stream in a straight annulus, and therefore conservation of mass requires that the limiting cross-sectional area of the cavity, A_C, be given by

$$\frac{A_C}{A_T} = 1 - \frac{U_\infty}{q_c} = 1 - (1 + \sigma_c)^{-\frac{1}{2}} \tag{8.12}$$

which leads to

$$\sigma_c \approx 2A_C/A_T \tag{8.13}$$

for the small values of the area ratio that would normally apply in water tunnel tests. The limiting cavity cross-sectional area, A_C, will be larger than the frontal body area, A_B. However, if the body is streamlined these areas will not differ greatly

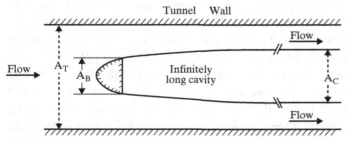

Figure 8.10 Body with infinitely long cavity under choked flow conditions.

and therefore, according to Equation (8.12), a first approximation to the value of σ_c would be

$$\sigma_c \approx 1 - \left(1 - \frac{A_B}{A_T}\right)^2 \tag{8.14}$$

Note, as could be anticipated, that the larger the blockage ratio, A_B/A_T, the higher the choked cavitation number, σ_c. Note, also, that the above equations assume frictionless flow since the relation, $q_c/U_\infty = (1+\sigma)^{\frac{1}{2}}$, was used. Hydraulic losses along the length of the water tunnel would introduce other effects in which choking would occur at the end of the tunnel working section in a manner analogous to the effects of friction in compressible pipe flow.

A second, useful result emerges when the momentum theorem is applied to the flow, again assumed frictionless. Then, in the limit of choked flow, the drag coefficient, $C_D(\sigma_c)$, is given by

$$C_D(\sigma_c) = \frac{A_T}{A_B}\left[\sigma_c - 2\left\{(1+\sigma_c)^{\frac{1}{2}} - 1\right\}\right] \tag{8.15}$$

When $\sigma_c \ll 1$ it follows from Equations (8.13) and (8.15) that

$$C_D(\sigma_c) \approx \frac{A_C^2}{A_B A_T} \approx \frac{A_C}{A_B}\frac{\sigma_c}{2} \tag{8.16}$$

where, of course, A_C/A_B, would depend on the shape of the body. The approximate validity of this result can be observed in Figure 8.16; it is clear that for the 30° half-angle wedge $A_C/A_B \approx 2$.

Wall effects and choked flow for lifting bodies have been studied by Cohen and Gilbert (1957), Cohen et al. (1957), Fabula (1964), Ai (1966), and others because of their importance to the water tunnel testing of hydrofoils. Moreover, similar phenomena will clearly occur in other internal flow geometries, for example that of a pump impeller. The choked cavitation numbers that emerge from such calculations can be very useful as indicators of the limiting cavitation operation of turbomachines such as pumps and turbines (see Section 8.9).

Finally, it is appropriate to add some comments on the wall effects in finite cavity flows for which $\sigma > \sigma_c$. It is counterintuitive that the blockage effect should cause a *reduction* in the drag at the same cavitation number as illustrated in Figure 8.16. Another remarkable feature of the wall effect, as Wu et al. (1971) demonstrate, is that the more streamlined the body the *larger* the fractional change in the drag caused by the wall effect. Consequently, it is *more* important to estimate and correct for the wall effects on streamline bodies than it is for bluff bodies with the same blockage ratio, A_B/A_T. Wu et al. (1971) evaluate these wall effects for the planar flows past cavitating wedges of various vertex angles (then $A_B/A_T = b/B$, figure 8.15) and suggest the following procedure for estimating the drag in the absence of wall effects. If during the experiment one were to measure the minimum coefficient of pressure, C_{pw}, on the tunnel wall at the point opposite the maximum width of the cavity, then Wu et al. recommend use of the following correction rule to estimate the coefficient of drag in the absence of wall effects, $C_D(\sigma',0)$, from the measured coefficient, $C_D(\sigma, b/B)$. The *effective* cavitation number for the unconfined flow is found to be σ' where

$$\sigma' = \sigma + 2C_{pw}(2-\sigma)/3(1-C_{pw}) \qquad (8.17)$$

and the unconfined drag coefficient is

$$C_D(\sigma', 0) = \frac{(1+\sigma')}{(1+\sigma)} C_D(\sigma, b/B) + O\left(\frac{b^2}{B^2}\right) \qquad (8.18)$$

As illustrated in Figure 8.16, Wu et al. (1971) use experimental data to show that this correction rule works well for flows around wedges with various vertex angles.

8.5 Steady Planar Flows

The classic free streamline solution for an arbitrary finite body with a fully developed cavity is obtained by mapping both the geometry of the physical plane (z-plane, Figure 8.2) and the geometry of the f-plane (Figure 8.11) into the lower half of a

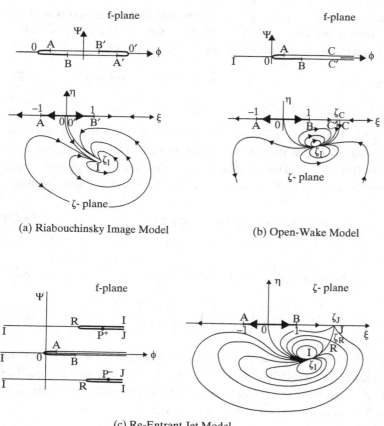

(a) Riabouchinsky Image Model   (b) Open-Wake Model

(c) Re-Entrant Jet Model

Figure 8.11 Streamlines in the complex potential f-plane and the parametric ζ-plane where the flow boundaries and points correspond to those of Figure 8.2.

parametric, ζ-plane. The wetted surface is mapped onto the interval, $\eta = 0$, $-1 < \xi < 1$ and the stagnation point, 0, is mapped into the origin. For the three closure models of Figure 8.2, the geometries of the corresponding ζ-planes are sketched in Figure 8.11. The $f = f(\zeta)$ mapping follows from the generalized Schwarz-Christoffel transformation (Gilbarg 1949); for the three closure models of Figures 8.2 and 8.11 this yields respectively

$$\frac{df}{d\zeta} = \frac{C\zeta}{(\zeta - \zeta_I)^{\frac{3}{2}}(\zeta - \bar{\zeta_I})^{\frac{3}{2}}} \tag{8.19}$$

$$\frac{df}{d\zeta} = \frac{C\zeta(\zeta - \zeta_C)}{(\zeta - \zeta_I)^2(\zeta - \bar{\zeta_I})^2} \tag{8.20}$$

$$\frac{df}{d\zeta} = \frac{C\zeta(\zeta - \zeta_R)(\zeta - \bar{\zeta_R})}{(\zeta - \zeta_I)^2(\zeta - \bar{\zeta_I})^2(\zeta - \zeta_J)} \tag{8.21}$$

where C is a real constant, ζ_I is the value of ζ at the point I (the point at infinity in the z-plane), ζ_C is the value of ζ at the end of the constant velocity part of the free streamlines, and ζ_R and ζ_J are the values at the rear stagnation point and the upstream infinity point in the reentrant jet model.

The wetted surface, AOB, will be given parametrically by $x(s), y(s)$ where s is the distance measured along that surface from the point A. Then the boundary conditions on the logarithmic hodograph variable, $\varpi = \chi + i\theta$, are

$$\theta^-(\xi) \equiv \theta(\xi, 0^-) = \pi F(-\xi) + \theta^*(s(\xi)) \quad \text{on} \quad -1 < \xi < 1 \tag{8.22}$$

$$\chi^-(\xi) \equiv \chi(\xi, 0^-) = 0 \quad \text{on} \quad \xi > 1 \quad \text{and} \quad \xi < -1 \tag{8.23}$$

where the superscripts $+$ and $-$ will be used to denote values on the ξ axis of the ζ-plane just above and just below the cut. The function $F(-\xi)$ takes a value of 1 for $\xi < 0$ and a value of 0 for $\xi > 0$. The function $\theta^*(s)$ is the inclination of the wetted surface so that $\tan\theta^* = dy/ds / dx/ds$. The solution to the above Reimann-Hilbert problem is

$$\varpi(\zeta) = \varpi_0(\zeta) + \varpi_1(\zeta) + \varpi_2(\zeta) \tag{8.24}$$

where

$$\varpi_0(\zeta) = \log\left\{\left(1 + i(\zeta^2 - 1)^{\frac{1}{2}}\right)/\zeta\right\} \tag{8.25}$$

$$\varpi_1(\zeta) = \frac{1}{i\pi}\int_{-1}^{1}\left(\frac{\zeta^2 - 1}{1 - \beta^2}\right)^{\frac{1}{2}}\frac{\theta^{**}(\beta)d\beta}{\beta - \zeta} \tag{8.26}$$

where β is a dummy variable, $\theta^{**}(\xi) = \theta^*(s(\xi))$ and the function $(\zeta^2 - 1)^{\frac{1}{2}}$ is analytic in the ζ-plane cut along the ξ axis from -1 to $+1$ so that it tends to ζ as $|\zeta| \to \infty$. The third function, $\varpi_2(\zeta)$, is zero for the Riabouchinsky and open-wake closure models; it is only required for the reentrant jet model and, in that case,

$$\varpi_2(\zeta) = \log\left\{\frac{(\beta - \bar{\beta}_R)(\beta\bar{\beta}_R - 1)}{(\beta - \beta_R)(\beta\bar{\beta}_R - 1)}\right\} \quad \text{where} \quad \zeta = (\beta + \beta^{-1})/2 \tag{8.27}$$

Given $\varpi(\zeta)$, the physical coordinate $z(\zeta)$ is then calculated using

$$z(\zeta) = \int \frac{1}{w(\zeta)} \frac{df}{d\zeta} d\zeta \tag{8.28}$$

The distance along the wetted surface from the point A is given by

$$s(\xi) = \frac{C}{q_c} \int_{-1}^{\xi} e^{\Gamma_1(\xi)} \Gamma_2(\xi) d\xi \tag{8.29}$$

where

$$\Gamma_1(\xi) = -\frac{1}{\pi} \oint_{-1}^{1} \left(\frac{1-\xi^2}{1-\beta^2} \right)^{\frac{1}{2}} \frac{\theta^{**}(\beta) d\beta}{\beta - \xi} \tag{8.30}$$

$$\Gamma_2(\xi) = \frac{1}{C} \exp\left\{ \varpi_0^-(\xi) + \varpi_2^-(\xi) \right\} \frac{df}{d\xi} \tag{8.31}$$

where the integral in Equation (8.30) takes its Cauchy principal value.

Now consider the conditions that can be applied to evaluate the unknown parameters in the problem, namely C and ζ_I in the case of the Riabouchinsky model, C, ζ_I, and ζ_C in the case of the open-wake model, and C, ζ_I, ζ_R, and ζ_J in the case of the reentrant jet model. All three models require that the total wetted surface length, $s(1)$, be equal to a known value, and this establishes the length scale in the flow. They also require that the velocity at $z \to \infty$ have the known magnitude, U_∞, and a given inclination, α, to the chord, AB. Consequently this condition becomes

$$\varpi(\zeta_0) = \frac{1}{2} \log(1+\sigma) + i\alpha \tag{8.32}$$

This is sufficient to determine the solution for the Riabouchinsky model. Additional conditions for the open-wake model can be derived from the fact that $f(\zeta)$ must be simply covered in the vicinity of ζ_0 and, for the reentrant jet model, that $z(\zeta)$ must be simply covered in the vicinity of ζ_0. Also the circulation around the cavity can be freely chosen in the re-entrant jet model. Finally, if the free streamline detachment is smooth and therefore initially unknown, its location must be established using the Brillouin-Villat condition (see Section 8.3). For further mathematical detail the reader is referred to the texts mentioned earlier in Section 8.1.

As is the case with all steady planar potential flows involving a body in an infinite uniform stream, the behavior of the complex velocity, $w(z)$, far from the body can be particularly revealing. If $w(z)$ is expanded in powers of $1/z$ then

$$w(z) = U_\infty e^{-i\alpha} + \frac{Q + i\Gamma}{2\pi} \frac{1}{z} + (C_1 + iC_2) \frac{1}{z^2} + O\left(\frac{1}{z^3}\right) \tag{8.33}$$

where U_∞ and α are the magnitude and inclination of the free stream. The quantity Q is the net source strength required to simulate the body-cavity system and must therefore be zero for a finite body-cavity. This constitutes a cavity closure condition. The quantity, Γ, is the circulation around the body-cavity so that the lift is given by

$\rho U_\infty \Gamma$. Evaluation of the $1/z$ term far from the body provides the simplest way to evaluate the lift.

The mathematical detail involved in producing results from these solutions (Wu and Wang 1964b) is considerable except for simple symmetric bodies. For more complex, bluff bodies it is probably more efficient to resort to one of the modern numerical methods (for example a panel method) rather than to attempt to sort through all the complex algebra of the above solutions. For streamlined bodies, a third alternative is the algebraically simpler linear theory for cavity flow, which is briefly reviewed in Section 8.7. There are, however, a number of valuable results that can be obtained from the above exact, nonlinear theory, and we will examine just a few of these in the next section.

8.6 Some Nonlinear Results

Wu (1956, 1962) (see also Mimura 1958) generated the solution for a flat plate at an arbitrary angle of incidence using the open-wake model and the methods described in the preceding section. The comparison between the predicted pressure distributions on the surface of the plate and those measured by Fage and Johansen (1927) in single phase, separated wake flow is excellent, as shown by the examples in Figure 8.12. Note that the effective cavitation number for the wake flow (or base pressure coefficient) is not an independent variable as it is with cavity flows. In Figure 8.12 the values of σ are taken from the experimental measurements. Data such as that presented in Figure 8.12 provides evidence that free streamline methods have value in wake flows as well as in cavity flows.

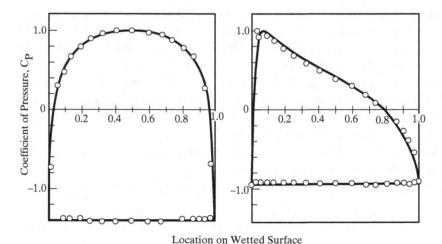

Figure 8.12 Comparison of pressure distributions on the surface of a flat plate set at an angle, α, to the oncoming stream. The theory of Wu (1956, 1962) (solid lines) is compared with the measurements in wake flow made by Fage and Johansen (1927) ($\circ$). The case on the left is for a flat plate set normal to the stream ($\alpha = 90°$) and a wake coefficient of $\sigma = 1.38$; the case on the right is $\alpha = 29.85°$, $\sigma = 0.924$. Adapted from Wu (1962).

The lift and drag coefficients at various cavitation numbers and angles of incidence are compared with the experimental data of Parkin (1958) and Silberman
(1959) in Figures 8.13 and 8.14. Data both for supercavitating and partially cavitating conditions are shown in these figures, the latter occurring at the higher cavitation
numbers and lower incidence angles. The calculations tend to be quite unstable in
the region of transition from the partially cavitating to the supercavitating state,
and so the dashed lines in Figures 8.13 and 8.14 represent smoothed curves in this
region. Later, in Section 8.8, we continue the discussion of this transition. For the
present, note that the nonlinear theory yields values for the lift and the drag that
are in good agreement with the experimental measurements. Wu and Wang (1964a)
show similar good agreement for supercavitating, circular-arc hydrofoils.

The solution to the cavity flow of a flat plate set normal to an oncoming stream,
$\alpha = 90°$, is frequently quoted (Birkhoff and Zarantonello 1957, Woods 1961), usually for the case of the Riabouchinsky model. At small cavitation numbers (large
cavities) the asymptotic form of the drag coefficient, C_D, is (Wu 1972)

$$C_D(\sigma) = \frac{2\pi}{\pi+4}\left[1 + \sigma + \frac{\sigma^2}{(8\pi+32)} + O(\sigma^4)\right] \tag{8.34}$$

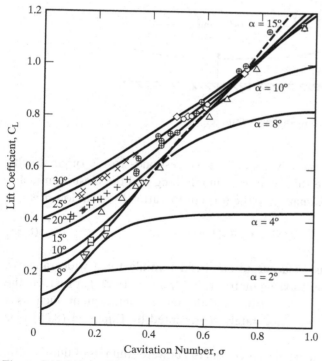

Figure 8.13 Lift coefficients for a flat plate from the nonlinear theory of Wu (1962). The
experimental data (Parkin 1958) is for angles of incidence as follows: 8° ($\triangledown$), 10° ($\square$), 15°
($\triangle$), 20° ($\oplus$), 25° ($\otimes$), and 30° ($\diamond$). Also shown is some data of Silberman (1959) in a free jet
tunnel: 20° ($+$) and 25° ($\times$).

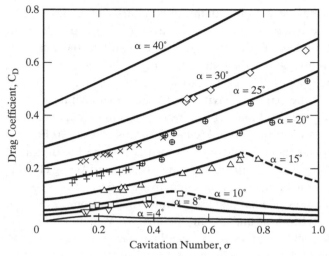

Figure 8.14 Drag coefficients corresponding to the lift coefficients of Figure 8.13.

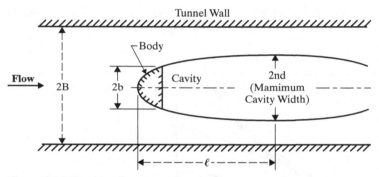

Figure 8.15 Notation for planar flow in a water tunnel.

where the value for $\sigma = 0$, namely $C_D = 0.88$, corresponds to the original solution of Kirchoff (in that case the cavity is infinitely long and the closure model is unnecessary). A good approximation to the form of Equation (8.34) at low σ is

$$C_D(\sigma) = C_D(0)[1 + \sigma] \qquad (8.35)$$

and it transpires that this is an accurate empirical formula for a wide range of body shapes, both planar and axisymmetric (see Brennen 1969a), provided the detachment is of the abrupt type. Bodies with smooth detachment such as a sphere (Brennen 1969a) are less accurately represented by Equation (8.35) (see Figure 8.18).

Since experiments are almost always conducted in water tunnels of finite width, $2B$, another set of solutions of interest are those in which straight tunnel boundaries are added to the geometries of the preceding section, as shown in Figure 8.15. In the case of symmetric wedges in tunnels, solutions for all three closure models of Figure 8.2 were obtained by Wu et al. (1971). Drag coefficients, cavity dimensions, and pressure distributions were computed as functions of cavitation number, σ, and blockage ratio, b/B. As illustrated in Figure 8.16, the results compare well with experimental measurements provided the cavitation number is low enough for a fully developed cavity to be formed (see Section 7.8). In the case shown in Figure 8.16, this cavitation number was about 1.5. The Riabouchinsky model results are shown in the figure since they were marginally better than those of the other two models insofar as the drag on the wedge was concerned. The variations with b/B shown in Figure 8.16 were discussed in Section 8.4.

For comparative purposes, some results for a cavitating sphere in an axisymmetric water tunnel are presented in Figures 8.17 and 8.18. These results were obtained by Brennen (1969a) using a numerical method (see Section 8.11). Note that the variations with tunnel blockage are qualitatively similar to those of planar flow. However, the calculated drag coefficients in Figure 8.18 are substantially larger than those experimentally measured because of the difference in the detachment locations discussed in Section 8.3 and illustrated in Figure 8.5.

Reichardt (1945) carried out some of the earliest experimental investigations of fully developed cavities and observed that, when the cavitation number becomes

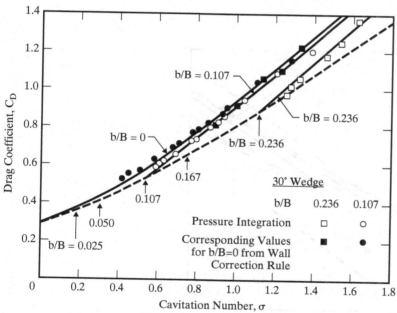

Figure 8.16 Analytical and experimental data for the drag coefficient, C_D, of a 30° half-angle wedge with a fully developed cavity in a water tunnel. Data are presented as a function of cavitation number, σ, for various ratios of wedge width to tunnel width, b/B (see Figure 8.15). Results are shown for the Riabouchinsky model (solid lines) including the choked flow conditions (dashed line with points for various b/B indicated by arrows), for the experimental measurements (open symbols), and for the experimental data corrected to $b/B \rightarrow 0$. Adapted from Wu, Whitney, and Brennen (1971).

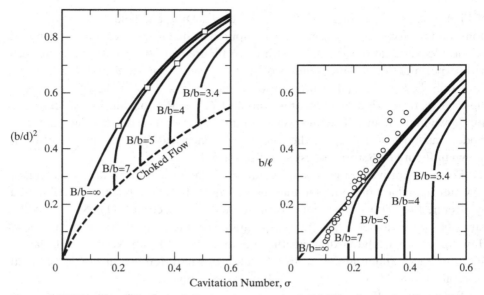

Figure 8.17 The dimensions of a fully developed cavity behind a sphere of radius, b, for various tunnel radii, B, from the numerical calculations of Brennen (1969a). On the left the maximum radius of the cavity, d, is compared with some results from Rouse and McNown (1948). On the right the half-length of the cavity, ℓ, is compared with the experimental data of Brennen (1969a) ($\circ$) for which $B/b = 14.7$.

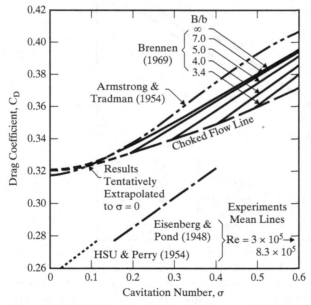

Figure 8.18 Calculated and measured drag coefficients for a sphere of radius, b, as a function of cavitation number, σ. The numerical results are by Armstrong and Tadman (1954) and Brennen (1969a) (for various tunnel radii, B) and the experimental data are from Eisenberg and Pond (1948) and Hsu and Perry (1954).

very small, the maximum width, $2d$, and the length, 2ℓ, of the cavity in an unconfined flow ($b/B = 0$) vary roughly with σ in the following way:

- In planar flow:

$$d \propto \sigma^{-1}; \quad \ell \propto \sigma^{-2} \tag{8.36}$$

- In axisymmetric flow:

$$d \propto \sigma^{-\frac{1}{2}}; \quad \ell \propto \sigma^{-1} \tag{8.37}$$

The data for $b/B = 0$ in Figure 8.17 are crudely consistent with the relations of Equation (8.37). Equations (8.36) and (8.37) provide a crude but useful guide to the relative dimensions of fully developed cavities at different cavitation numbers.

8.7 Linearized Methods

When the body/cavity system is slender in the sense that the direction of the velocity vector is everywhere close to that of the oncoming uniform stream (except, perhaps, close to some singularities), then methods similar to those of thin airfoil theory (see, for example, Biot 1942) become feasible. The approximations involved lead to a more tractable mathematical problem and to approximate solutions in circumstances in which the only alternative would be the application of more direct numerical methods. Linear theories for cavity flows were pioneered by Tulin (1953). Though the methods have been extended to three-dimensional flows, it is convenient to begin by describing their application to the case of an inviscid and incompressible planar flow of a uniform stream of velocity, U_∞, past a single, streamlined cavitating body. It is assumed that the body is slender and that the wetted surface is described by $y = h(x)$ where $dh/dx \ll 1$. It is also assumed that the boundary conditions on the body and the cavity can, to a first approximation, be applied on the x-axis as shown in Figure 8.19. The velocity components at any point are denoted by $u = U_\infty + u'$ and v where the linearization requires that both u' and v are much smaller than U_∞. The appropriate boundary condition on the wetted

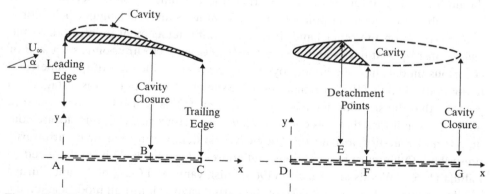

Figure 8.19 Examples of the linearized geometry (lower figures) for two planar cavity flows (upper sketches): a partially cavitating foil (left) and a supercavitating headform (right). Solid boundaries are indicated by the thick lines and the free streamlines by the thick dashed lines.

surface is then

$$v = U_\infty \frac{dh}{dx} \tag{8.38}$$

Moreover, the coefficient of pressure anywhere in the flow is given by $C_p \approx -2u'/U_\infty$, and therefore the boundary condition on a free streamline becomes

$$u' = \sigma U_\infty/2 \tag{8.39}$$

Finally, a boundary condition at infinity must also be prescribed. In some instances it seems appropriate to linearize about an x-axis that is parallel with the velocity at infinity. In other cases, it may be more appropriate and more convenient to linearize about an x-axis that is parallel with a mean longitudinal line through the body-cavity system. In the latter case the boundary condition at infinity is $w(z \to \infty) \to U_\infty e^{-i\alpha}$ where α is the angle of incidence of the uniform stream relative to the body-cavity axis.

Even within the confines of this simple problem, several different configurations of wetted surface and free surface are possible, as illustrated by the two examples in Figure 8.19. Moreover, various types of singularity can occur at the end points of any segment of boundary in the linearized plane (points A through G in Figure 8.19). It is important that the solution contain the correct singular behavior at each of these points. Consider the form that the complex conjugate perturbation velocity, $w = u' - iv$, must take for each of the different types of singularity that can occur. Let $x = c$ be the location of the specific singularity under consideration. Clearly, then, a point like D, the stagnation point at a rounded nose or leading edge, must have a solution of the form $w \sim i(z - c)^{-1/2}$ (Newman 1956). On the other hand, a sharp leading edge from which a free surface detaches (such as A) must have the form $w \sim i(z - c)^{-1/4}$ (Tulin 1953). These results are readily derived by applying the appropriate conditions of constant v or constant u' on $\theta = 0$ and $\theta = 2\pi$.

The conditions at regular detachment points such as E or F (as opposed to the irregular combination of a detachment point and front stagnation point at A) should follow the conditions derived earlier for detachment points (Section 8.3). If it is an abrupt detachment point, then w is continous and the singular behavior is $w \sim (z - c)^{1/2}$; on the other hand, if it is a smooth detachment point, both w and dw/dz must be continuous and $w \sim (z - c)^{3/2}$. At cavity closure points such as B or G various models have been employed (Tulin 1964). In the case of the supercavitating body, Tulin's (1953) original model assumes that the point G is a stagnation point so that the singular behavior is $w \sim (z - c)^{1/2}$; this is also the obvious choice under the conditions that u' is constant on $\theta = \pm\pi$. However, with this closure condition the circulation around the body-cavity system can no longer be arbitrarily prescribed. Other closure conditions that address this issue have been discussed by Fabula (1962), Woods and Buxton (1966), Nishiyama and Ota (1971), and Furuya (1975a), among others. In the case of the partial cavity almost all models assume a stagnation point at the point B so that the singular behavior is $w \sim (z - c)^{1/2}$. The problem of prescription of circulation that occurred with the supercavitation closure does not arise in this case since the conventional, noncavitating Kutta condition can be applied at the trailing edge, C.

The literature on linearized solutions for cavity flow problems is too large for thorough coverage in this text, but a few important milestones should be mentioned. Tulin's (1953) original work included the solution for a supercavitating flat plate hydrofoil with a sharp leading edge. Shortly thereafter, Newman (1956) showed how a rounded leading edge might be incorporated into the linear solution and Cohen, Sutherland, and Tu (1957) provided information on the wall effects in a tunnel of finite width. Acosta (1955) provided the first partial cavitation solution, specifically for a flat plate hydrofoil (see below). For a more recent treatment of supercavitating single foils, the reader is referred to Furuya and Acosta (1973).

It is appropriate to examine the linear solution to a typical cavity flow problem and, in the next section, the details for a cavitating flat plate hydrofoil will be given.

Many other types of cavitating flow have been treated by linear theory, including such problems as the effect of a nearby ocean surface. An important class of solutions is that involving cascades of foils, and these are addressed in Section 8.9.

8.8 Flat Plate Hydrofoil

The algebra associated with the linear solutions for a flat plate hydrofoil is fairly simple, so we will review and examine the results for the supercavitating foil (Tulin 1953) and for the partially cavitating foil (Acosta 1955). Starting with the latter, the z-plane is shown on the left in Figure 8.19, and this can be mapped into the upper half of the ζ-plane in Figure 8.20 by

$$\zeta = i \left(1 + \frac{1}{z-1} \right)^{\frac{1}{2}} \tag{8.40}$$

The point H_∞ at $\eta = i$ corresponds to the point at infinity in the z-plane and the point C_∞, the trailing edge of the foil, is the point at infinity in the ζ-plane. It follows that the point B, the cavity closure point, is at $\xi = c$ where $c = (\ell/(1-\ell))^{1/2}$ and ℓ is the length of the cavity, AB, in the physical plane. The chord of the hydrofoil, AC, has been set to unity. Since there must be square-root singularities at A and B, since v is zero on the real axis in the intervals $\xi < 0$ and $\xi > c$ and $u' = \sigma U_\infty/2$ in $0 < \xi < c$, and since w must be everywhere bounded, the general form of the solution may be

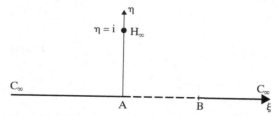

Figure 8.20 The ζ-plane for the linearized theory of a partially cavitating flat plate hydrofoil.

written down by inspection:

$$w(\zeta) = \frac{\sigma U_\infty}{2} + \frac{U_\infty(C_0 + C_1\zeta)}{[\zeta(\zeta - c)]^{1/2}} \tag{8.41}$$

where C_0 and C_1 are constants to be determined. The Kutta condition at the trailing edge, C_∞, requires that the velocity be finite and continous at that point, and this is satisfied provided there are no terms of order ζ^2 or higher in the series $C_0 + C_1\zeta$.

The conditions that remain to be applied are those at the point of infinity in the physical plane, $\eta = i$. The nature of the solution near this point should therefore be examined by expanding in powers of $1/z$. Since $\zeta \to i + i/2z + O(z^{-2})$ and since we must have that $w \to -i\alpha U_\infty$, expanding Equation (8.41) in powers of $1/z$ allows evaluation of the real constants, C_0 and C_1, in terms of α and σ:

$$C_0 = \beta_1\alpha + \beta_2\sigma/2; \quad C_1 = \beta_2\alpha - \beta_1\sigma/2 \tag{8.42}$$

where

$$\beta_1, \beta_2 = \left[\frac{1}{2}\left\{ (1-\ell)^{-1/2} \pm 1 \right\} \right]^{\frac{1}{2}} \tag{8.43}$$

In addition, the expansion of w in powers of $1/z$ must satisfy Equation (8.33). If the cavity is finite, then $Q = 0$ and evaluation of the real part of the coefficient of $1/z$ leads to

$$\frac{\sigma}{2\alpha} = \frac{2 - \ell + 2(1-\ell)^{\frac{1}{2}}}{\ell^{\frac{1}{2}}(1-\ell)^{\frac{1}{2}}} \tag{8.44}$$

while the imaginary part of the coefficient of $1/z$ allows the circulation around the foil to be determined. This yields the lift coefficient,

$$C_L = \pi\alpha \left[1 + (1-\ell)^{-\frac{1}{2}} \right] \tag{8.45}$$

Thus the solution has been obtained in terms of the parameter, ℓ, the ratio of the cavity length to the chord. For a given value of ℓ and a given angle of attack, α, the corresponding cavitation number follows from Equation (8.44) and the lift coefficient from equation 8.45. Note that as $\ell \to 0$ the value of C_L tends to the theoretical value for a noncavitating flat plate, $2\pi\alpha$. Also note that the lift-slope, $dC_L/d\alpha$, tends to infinity when $\ell = 3/4$.

In the supercavitating case, Tulin's (1953) solution yields the following results in place of Equations (8.44) and (8.45):

$$\alpha\left(\frac{2}{\sigma} + 1\right) = (\ell - 1)^{\frac{1}{2}} \tag{8.46}$$

$$C_L = \pi\alpha\ell \left[\ell^{\frac{1}{2}}(\ell - 1)^{-\frac{1}{2}} - 1 \right] \tag{8.47}$$

where now, of course, $\ell > 1$. Note that the lift-slope, $dC_L/d\alpha$, is zero at $\ell = 4/3$.

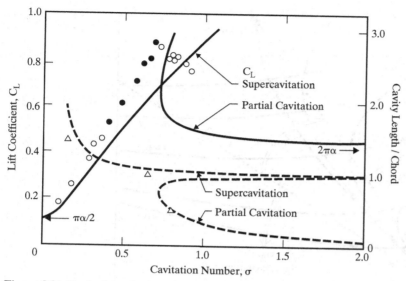

Figure 8.21 Typical results from the linearized theories for a cavitating flat plate at an angle of attack of 4°. The lift coefficients, C_L (solid lines), and the ratios of cavity length to chord, ℓ (dashed lines), are from the supercavitation theory of Tulin (1953) and the partial cavitation theory of Acosta (1955). Also shown are the experimental results of Wade and Acosta (1966) for ℓ ($\triangle$) and for C_L (○ and ●) where the open symbols represent points of stable operation and the solid symbols denote points of unstable cavity operation.

The lift coefficient and the cavity length from Equations (8.44) to (8.47) are plotted against cavitation number in Figure 8.21 for a typical angle of attack of $\alpha = 4°$. Note that as $\sigma \to \infty$ the fully wetted lift coefficient, $2\pi\alpha$, is recovered from the partial cavitation solution, and that as $\sigma \to 0$ the lift coefficient tends to $\pi\alpha/2$. Notice also that both the solutions become pathological when the length of the cavity approaches the chord length ($\ell \to 1$). However, if some small portion of each curve close to $\ell = 1$ is eliminated, then the characteristic decline in the performance of the hydrofoil as the cavitation number is decreased can be observed. Specifically, it is seen that the decline in the lift coefficient begins when σ falls below about 0.7 for the flat plate at an angle of attack of 4°. Close to $\sigma = 0.7$, one observes a small increase in C_L before the decline sets in, and this phenomenon is often observed in practice, as illustrated by the experimental data of Wade and Acosta (1966) included in Figure 8.21.

The variation in the lift with angle of attack (for a fixed cavitation number) is presented in Figure 8.22. Also shown in this figure are the lines of $\ell = 4/3$ in the supercavitation solution and $\ell = 3/4$ in the partial cavitation solution. Note that these lines separate regions for which $dC_L/d\alpha > 0$ from those for which $dC_L/d\alpha < 0$. Heuristically it could be argued that $dC_L/d\alpha < 0$ implies an unstable flow and the corresponding region in figure 8.22 for which $3/4 < \ell < 4/3$ does, indeed, correspond quite closely to the observed regime of unstable cavity oscillation (Wade and Acosta 1966).

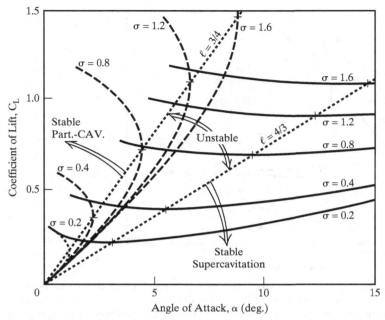

Figure 8.22 The lift coefficient for a flat plate from the partial cavitation model of Acosta (1955) (dashed lines) and the supercavitation model of Tulin (1953) (solid lines) as a function of angle of attack, α, for several cavitation numbers, σ, as shown. The dotted lines are the boundaries of the region in which the cavity length is between 3/4 and 4/3 of a chord and in which $dC_L/d\alpha < 0$.

8.9 Cavitating Cascades

Because cavitation problems are commonly encountered in liquid turbomachines (pumps, turbines) and on propellers, the performance of a cascade of hydrofoils under cavitating conditions is of considerable practical importance. A typical cascade geometry (z-plane) is shown on the left in Figure 8.23; in the terminology of these flows the angle, β, is known as the "stagger angle" and $1/h$, the ratio of the blade chord to the distance between the blade tips, is known as the "solidity." The corresponding complex potential plane (f-plane) is shown on the right. Note that the geometry of the linearized physical plane is very similar to that of the f-plane.

The first step in the analysis of the planar potential flow in a cascade (whether by linear or nonlinear methods) is to map the infinite array of blades in the f-plane (or the linearized z-plane) into a ζ-plane in which there is a single wetted surface boundary and a single cavity surface boundary. This is accomplished by the well-known cascade mapping function

$$f \quad \text{or} \quad z = \frac{h}{2\pi} \left[e^{-i\beta} \ln(1 - \zeta/\zeta_H) + e^{i\beta} \ln(1 - \zeta/\bar{\zeta}_H) \right] \tag{8.48}$$

where h^* (or h) is the distance between the leading edges of the blades and β^* (or β) is the stagger angle of the cascade in the f-plane (or the linearized z-plane). This mapping produces the ζ-plane shown in Figure 8.24 where H_∞

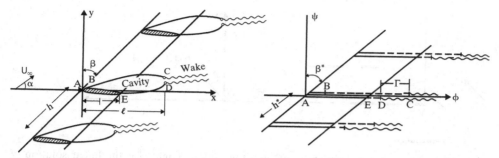

Figure 8.23 On the left is the physical plane (z-plane), and on the right is the complex potential plane (f-plane) for the planar flow through a cascade of cavitating hydrofoils. The example shown is for supercavitating foils. For partial cavitation the points D and E merge and the point C is on the upper wetted surface of the foil. For a sharp leading edge the points A and B merge. Figures adapted from Furuya (1975a).

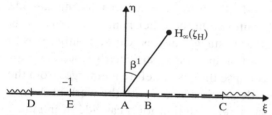

Figure 8.24 The ζ-plane obtained by using the cascade mapping function.

($\zeta = \zeta_H$) is the point at infinity in the original plane and the angle β' is equal to the stagger angle in the original plane. The solution is obtained when the mapping $w(\zeta)$ has been determined and all the boundary conditions have been applied.

For a supercavitating cascade, a nonlinear solution was first obtained by Woods and Buxton (1966) for the case of a cascade of flat plates. Furuya (1975a) expanded this work to include foils of arbitrary geometry. An interesting innovation introduced by Woods and Buxton was the use of Tulin's (1964) double-spiral-vortex model for cavity closure, but with the additional condition that the difference in the velocity potentials at the points C and D (Figure 8.23) should be equal to the circulation around the foil.

Linear theories for a cascade began much earlier with the work of Betz and Petersohn (1931), who solved the problem of infinitely long, open cavities produced by a cascade of flat plate hydrofoils. Sutherland and Cohen (1958) generalized this to the case of finite supercavities behind a flat plate cascade, and Acosta (1960) solved the same problem but with a cascade of circular-arc hydrofoils. Other early contributions to linear cascade theory for supercavitating foils include the models of Duller (1966) and Hsu (1972) and the inclusion of the effect of rounded leading edges by Furuya (1974).

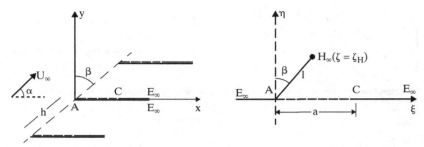

Figure 8.25 The linearized z-plane (left) and the ζ-plane (right) for the linear solution of partial cavitation in an infinitely long cascade of flat plates (Acosta and Hollander 1959). The points E_∞ and H_∞ are respectively the points at upstream and downstream infinity in the z-plane.

Cavities initiated at the leading edge are more likely to extend beyond the trailing edge when the solidity and the stagger angle are small. Such cascade geometries are more characteristic of propellers and, therefore, the supercavitating cascade results are more often utilized in that context. On the other hand, most cavitating pumps have larger solidities (> 1) and large stagger angles. Consequently, partial cavitation is the more characteristic condition in pumps, particularly since the pressure rise through the pump is likely to collapse the cavity before it emerges from the blade passage. Partially cavitating cascade analysis began with the work of Acosta and Hollander (1959), who obtained the linear solution for a cascade of infinitely long flat plates, the geometry of which is shown in Figure 8.25. The appropriate cascade mapping is then the version of Equation (8.48) with z on the left-hand side. The Acosta and Hollander solution is algebraically simple and therefore makes a good, specific example. The length of the cavity in the ζ-plane, a, provides a convenient parameter for the problem and should not be confused with the actual cavity length, ℓ. Given the square-root singularities at A and C, the complex velocity, $w(\zeta)$, takes the form

$$\frac{w(\zeta)}{U_\infty} = (1+\sigma)^{\frac{1}{2}} + C_1 \left[\frac{\zeta}{\zeta-a}\right]^{\frac{1}{2}} - C_2 \left[\frac{\zeta-a}{\zeta}\right]^{\frac{1}{2}} \tag{8.49}$$

where the real constants, C_1 and C_2, must be determined by the conditions at upstream and downstream infinity. As $x \to -\infty$ or $\zeta \to \zeta_H = ie^{-i\beta}$ we must have $w/U_\infty = e^{-i\alpha}$ and therefore

$$C_1 \left[\frac{\zeta_H}{\zeta_H - a}\right]^{\frac{1}{2}} - C_2 \left[\frac{\zeta_H - a}{\zeta_H}\right]^{\frac{1}{2}} + (1+\sigma)^{\frac{1}{2}} = e^{-i\alpha} \tag{8.50}$$

and, as $x \to +\infty$ or $|\zeta| \to \infty$, continuity requires that

$$C_1 - C_2 + (1+\sigma)^{\frac{1}{2}} = \cos(\alpha+\beta)/\cos(\beta) \tag{8.51}$$

The complex Equation (8.50) and the scalar Equation (8.51) permit evaluation of C_1, C_2, and a in terms of the parameters of the physical problem, σ and β. Then the completed solution can be used to evaluate such features as the cavity length, ℓ:

$$\frac{\ell}{h} = \frac{1}{\pi} Re \left\{ e^{-i\beta} \ln \left(1 + iae^{i\beta} \right) \right\} \tag{8.52}$$

Wade (1967) extended this partial cavitation analysis to cover flat plate foils of finite length, and Stripling and Acosta (1962) considered the nonlinear problem. Brennen and Acosta (1973) presented a simple, approximate method by which a finite blade thickness can be incorporated into the analysis of Acosta and Hollander. This is particularly valuable because the choked cavitation number, σ_c, is quite sensitive to the blade thickness or radius of curvature of the leading edge. The following is the expression for σ_c from the Brennen and Acosta analysis:

$$\sigma_c = \left[1 + 2\sin\frac{\alpha}{2}\sec\left(\frac{\pi}{4} - \frac{\beta}{2}\right)\sin\left(\frac{\pi}{4} - \frac{\beta}{2} - \frac{\alpha}{2}\right) + 2d\sin^2\left(\frac{\pi}{4} - \frac{\beta}{2}\right) \right]^{\frac{1}{2}} - 1 \tag{8.53}$$

where d is the ratio of the blade thickness to normal blade spacing, $h\cos\beta$, far downstream. Since the validity of the linear theory requires that $\alpha \ll 1$ and since many pumps (for example, cavitating inducers) have stagger angles close to $\pi/2$, a reasonable approximation to Equation (8.53) is

$$\sigma_c \approx \alpha(\theta - \alpha) + \theta^2 d \tag{8.54}$$

where $\theta = \pi/2 - \beta$. This limit is often used to estimate the breakdown cavitation number for a pump based on the heuristic argument that long partial cavities that reach the pump discharge would permit substantial deviation angles and therefore lead to a marked decline in pump performance (Brennen and Acosta 1973).

Note, however, that under the conditions of an inviscid model, a small partial cavity will not significantly alter the performance of the cascade of higher solidity (say, $1/h > 1$) since the discharge, with or without the cavity, is essentially constrained to follow the direction of the blades. On the other hand, the direction of flow downstream of a supercavitating cascade will be significantly affected by the cavities, and the corresponding lift and drag coefficients will be altered by the cavitation. We return to the subject of supercavitating cascades to demonstrate this effect.

A substantial body of data on the performance of cavitating cascades has been accumulated through the efforts of Numachi (1961, 1964), Wade and Acosta (1967), and others. This allows comparison with the analytical models, in particular the supercavitating theories. Figure 8.26 provides such a comparison between measured lift and drag coefficients (defined as normal and parallel to the direction of the incident stream) for a particular cascade and the theoretical results from the supercavitating theories of Furuya (1975a) and Duller (1966). Note that the measured lift coefficients exhibit a rapid decline in cascade performance as the cavitation number is reduced and the supercavities grow. However, it is important to observe that

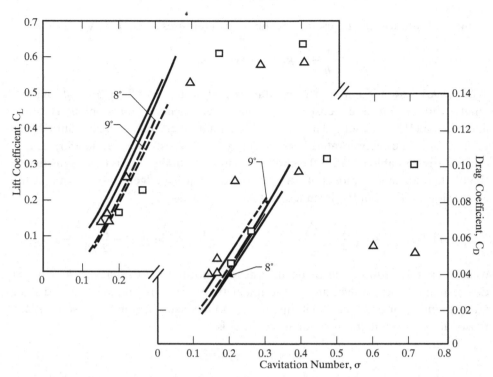

Figure 8.26 Lift and drag coefficients as functions of the cavitation number for cascades of solidity, 0.625, and stagger angle, $\beta = 45° - \alpha$, operating at angles of incidence, α, of 8° ($\triangle$) and 9° ($\square$). The points are from the experiments of Wade and Acosta (1967), and the analytical results for a supercavitating cascade are from the linear theory of Duller (1966) (dashed lines) and the nonlinear theory of Furuya (1975a) (solid lines).

this degradation does not occur until the cavitation is quite extensive. The cavitation inception numbers for the experiments were $\sigma_i = 2.35$ (for 8°) and $\sigma_i = 1.77$ (for 9°). However, the cavitation number must be lowered to about 0.5 before the performance is adversely affected. In the range of σ in between are the partial cavitation states for which the performance is little changed.

For the cascades and incidence angles used in the example of Figure 8.26, Furuya (1975a) shows that the linear and nonlinear supercavitation theories yield results that are similar and close to those of the experiments. This is illustrated in Figure 8.26. However, Furuya also demonstrates that there are circumstances in which the linear theories can be substantially in error and for which the nonlinear results are clearly needed. The effect of the solidity, $1/h$, on the results is also important because it is a major design factor in determining the number of blades in a pump or propeller. Figure 8.27 illustrates the effect of solidity when large supercavities are present ($\sigma = 0.18$). Note that the solidity has remarkably little effect.

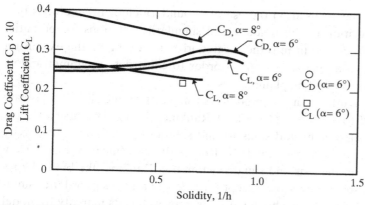

Figure 8.27 Lift and drag coefficients as functions of the solidity for cascades of stagger angle, $\beta = 45° - \alpha$, operating at the indicated angles of incidence, α, and at a cavitation number, $\sigma = 0.18$. The points are from the experiments of Wade and Acosta (1967), and the lines are from the nonlinear theory of Furuya (1975). Reproduced from Furuya (1975a).

8.10 Three-Dimensional Flows

Though numerical methods seem to be in the ascendant, several efforts have been made to treat three-dimensional cavity flows analytically. Early analyses of attached cavities on finite aspect ratio foils combined the solutions for planar flows with the corrections known from finite aspect ratio aerodynamics (Johnson 1961). Later, stripwise solutions for cavitating foils of finite span were developed in which an inner solution from either a linear or a nonlinear theory was matched to an outer solution from lifting line theory. This approach was used by Nishiyama (1970), Leehey (1971), and Furuya (1975b) to treat supercavitating foils and by Uhlman (1978) for partially cavitating foils. Widnall (1966) used a lifting surface method in a three-dimensional analysis of supercavitating foils.

For more slender bodies such as delta wings, the linearized procedure outlined in Section 8.7 can be extended to three-dimensional bodies in much the same way as it is applied in the slender body theories of aerodynamics. Tulin (1959) and Cumberbatch and Wu (1961) used this approach to model cavitating delta wings.

8.11 Numerical Methods

With the modern evolution of computational methods it has become increasingly viable to consider more direct numerical methods for the solution of free surface flows, even in circumstances in which analytical solutions could be generated. It would be beyond the scope of this text to survey these computational methods, and so we confine our discussion to some brief comments on the methods used in the past. These can be conveniently divided into two types. Some of the literature describes "field" methods in which the entire flow field is covered by a lattice of grids and node points at which the flow variables are evaluated. But most of the work in the past has focused on the use of "boundary element" methods that make use of superposition of the fundamental singularity solutions for potential flows. A few methods do not fit into these categories; for example, the expansion technique devised by Garabedian (1956) in order to construct axisymmetric flow solutions from the corresponding planar flows.

Methods for the synthesis of potential flows using distributed singularities can, of course, be traced to the original work of Rankine (1871). The first attempts to use distributions of sources and sinks to find solutions to axisymmetric cavity flow problems appear to have been made by Reichardt and Munzner (1950). They distributed doublets on the axis and sought symmetric, Rankine-like body shapes with nearly constant surface pressure except for fore and aft caps in order to simulate Riabouchinsky flows. The problem with this approach is its inability to model the discontinuous or singular behavior at the free surface detachment points. This requires a distribution of surface singularities that can either be implemented explicitly (most conveniently with surface vortex sheet elements) or by the equivalent use of Green's function methods as pioneered by Trefftz (1916) in the context of jets. Distributions of surface singularities to model cavity flows were first employed by Landweber (1951), Armstrong and Dunham (1953), and Armstrong and Tadman (1954). The latter used these methods to generate solutions for the axisymmetric Riabouchinsky solutions of cavitating discs and spheres. The methods were later extended to three-dimensional potential flows by Struck (1966), who addressed the problem of an axisymmetric body at a small angle of attack to the oncoming stream.

As computational capacity grew, it became possible to examine more complex three-dimensional flows and lifting bodies using boundary element methods. For example, Lemonnier and Rowe (1988) computed solutions for a partially cavitating hydrofoil and Uhlman (1987, 1989) has generated solutions for hydrofoils with both partial cavitation and supercavitation. These methods solve for the velocity. The position of the cavity boundary is determined by an iterative process in which the dynamic condition is satisfied on an approximate cavity surface and the kinematic condition is used to update the location of the surface. More recently, a method that uses Green's theorem to solve for the potential has been developed by Kinnas and Fine (1990) and has been applied to both partially and supercavitating hydrofoils. This appears to be superior to the velocity-based methods in terms of convergence.

Efforts have also been made to develop "field" methods for cavity flows. Southwell and Vaisey (1946) (see also Southwell 1948) first explored the use of relaxation methods to solve free surface problems but did not produce solutions for any realistic cavity flows. Woods (1951) suggested that solutions to axisymmetric cavity flows could be more readily obtained in the geometrically simpler (ϕ, ψ) plane, and Brennen (1969a) used this suggestion to generate Riabouchinsky model solutions for a cavitating disc and sphere in a finite water tunnel (see Figures 8.5, 8.17 and 8.18). In more recent times, it has become clear that boundary integral methods are more efficient for potential flows. However, field methods must still be used when seeking solutions to the more complete viscous flow problem. Significant progress has been made in the last few years in developing Navier-Stokes solvers for free surface problems in general and cavity flow problems in particular (see, for example, Deshpande et al. 1993).

8.12 Unsteady Flows

Most of the analyses in the preceding sections addressed various *steady* free stream-line flows. The corresponding unsteady flows pose more formidable modeling problems, and it is therefore not surprising that progress in solving these unsteady flows has been quite limited. Though Wang and Wu (1965) show how a general perturbation theory of cavity flows may be formulated, the implementation of their methodology to all but the simplest flows may be prohibitively complicated. More-over, there remains much uncertainty regarding the appropriate closure model to use in unsteady flow. Consequently, the case of zero cavitation number raises less uncertainty since it involves an infinitely long cavity and no closure. We will therefore concentrate on the linear solution of the problem of small amplitude per-turbations to a mean flow with zero cavitation number. This problem was first solved by Woods (1957) in the context of an oscillating aerofoil with separated flow but can be more confidently applied to the cavity flow problem. Martin (1962) and Parkin (1962) further refined Woods' theory and provided tabulated data for the unsteady force coefficients, which we will utilize in this summary.

The unsteady flow problem is best posed using the "acceleration potential" (see, for example, Biot 1942), denoted here by ϕ' and defined simply as $(p_\infty - p)/\rho$, so that linearized versions of Euler's equations of motion may be written as

$$\frac{\partial \phi'}{\partial x} = \frac{\partial u}{\partial t} + U_\infty \frac{\partial u}{\partial x} \tag{8.55}$$

$$\frac{\partial \phi'}{\partial y} = \frac{\partial v}{\partial t} + U_\infty \frac{\partial v}{\partial x} \tag{8.56}$$

It follows from the equation of continuity that ϕ' satisfies Laplace's equation,

$$\nabla^2 \phi' = 0 \tag{8.57}$$

Now consider the boundary conditions on the cavity and on the wetted surface of a flat plate foil. Since the cavity pressure at zero cavitation number is equal to p_∞, it follows that the boundary condition on a free surface is $\phi' = 0$. The linearized condition on a wetted surface (the unsteady version of Equation (8.38)) is clearly

$$v = -\alpha U_\infty - \frac{\partial h}{\partial t} \tag{8.58}$$

where $y = -h(x,t)$ describes the geometry of the wetted surface, α is the angle of incidence, and the chord of the foil is taken to be unity. We consider a flat plate at a mean angle of incidence of $\bar{\alpha}$ that is undergoing small-amplitude oscillations in both heave and pitch at a frequency, ω. The amplitude and phase of the pitching oscillations are incorporated in the complex quantity, $\tilde{\alpha}$, so that the instantaneous angle of incidence is given by

$$\alpha = \bar{\alpha} + Re\left\{\tilde{\alpha}e^{j\omega t}\right\} \qquad (8.59)$$

and the amplitude and phase of the heave oscillations of the leading edge are incorporated in the complex quantity $\tilde{h}$ (positive in the negative y direction) so that

$$h(x,t) = Re\left\{\tilde{h}e^{j\omega t}\right\} + xRe\left\{\tilde{\alpha}e^{j\omega t}\right\} \qquad (8.60)$$

where the origin of x is taken to be the leading edge. Combining Equations (8.56), (8.58), and (8.60), the boundary condition on the wetted surface becomes

$$\frac{\partial \phi'}{\partial y} = Re\left\{(\omega^2\tilde{h} + \omega^2\tilde{\alpha}x - 2j\omega U_\infty\tilde{\alpha})e^{j\omega t}\right\} \qquad (8.61)$$

Consequently, the problem reduces to solving for the analytic function $\phi'(z)$ subject to the conditions that ϕ' is zero on a free streamline and that, on a wetted surface, $\partial\phi'/\partial y$ is a known, linear function of x given by Equation (8.61).

In the linearized form this mathematical problem is quite similar to that of the steady flow for a cavitating foil at an angle of attack and can be solved by similar methods (Woods 1957, Martin 1962). The resulting instantaneous lift and moment coefficients can be decomposed into components due to the pitch and the heave:

$$C_L = \bar{C}_L + Re\left\{(\tilde{h}\tilde{C}_{Lh} + \tilde{\alpha}\tilde{C}_{Lp})e^{j\omega t}\right\} \qquad (8.62)$$

$$C_M = \bar{C}_M + Re\left\{(\tilde{h}\tilde{C}_{Mh} + \tilde{\alpha}\tilde{C}_{Mp})e^{j\omega t}\right\} \qquad (8.63)$$

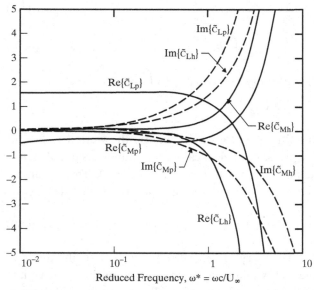

Figure 8.28 Real and imaginary parts of the four unsteady lift and moment coefficients for a flat plate hydrofoil at zero cavitation number.

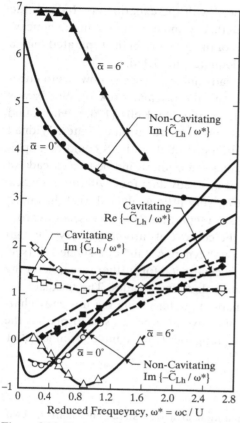

Figure 8.29 Fluctuating lift coefficients, $\tilde{C}_{Lh}$, for foils undergoing heave oscillations at various reduced frequencies, ω^*. Real and imaginary parts of $\tilde{C}_{Lh}/\omega*$ are presented for noncavitating flow at mean incidence angles of $0°$ and $6°$ (solid symbols) and for cavitating flow for a mean incidence of $8°$, for very long choked cavities ($\square$) and for cavities 3 chords in length ($\diamond$). Adapted from Acosta and DeLong (1971).

where the moment about the leading edge is considered positive in the clockwise direction (tending to increase α). The four complex coefficients, $\tilde{C}_{Lh}$, $\tilde{C}_{Lp}$, $\tilde{C}_{Mh}$, and $\tilde{C}_{Mp}$, represent the important dynamic characteristics of the foil and are functions of the reduced frequency defined as $\omega^* = \omega c/U_\infty$ where c is the chord. The tabulations by Parkin (1962) allow evaluation of these coefficients, and they are presented in Figure 8.28 as functions of the reduced frequency. The values tabulated by Woods (1957) yield very similar results. Note that when the reduced frequency is much less than unity, the coefficients tend to their quasistatic values; in this limit all but $Re\{\tilde{C}_{Lp}\}$ and $Re\{\tilde{C}_{Mp}\}$ tend to zero, and these two nonzero coefficients tend to the quasistatic values of $dC_L/d\alpha$ and $dC_M/d\alpha$, namely $\pi/2$ and $5\pi/32$, respectively.

Acosta and DeLong (1971) measured the oscillating forces on a cavitating hydrofoil subjected to heave oscillations at various reduced frequencies. Their results both for cavitating and noncavitating flow are presented in Figure 8.29 for several mean angles of incidence, $\bar{\alpha}$. The analytical results from figure 8.28 are included in this figure and compare fairly well with the experiments. Indeed, the agreement is better than is manifest between theory and experiment in the noncavitating case, perhaps because the oscillations of the pressure in the separated region or wake of the noncavitating flow are not adequately modeled.

Other advances in the treatment of unsteady linearized cavity flows were introduced by Wu (1957) and Timman (1958), and the original work of Woods was extended to finite cavitation numbers (finite cavities) by Kelly (1967), who found that the qualitative nature of the results was not dependent on σ. Later, Widnall (1966) showed how the linearized acceleration potential methods could be implemented in three dimensions. Another valuable extension would be to a cascade of foils, but the author is unaware of any similar unsteady data for cavitating cascades. Indeed, apart from the work of Sisto (1967), very little analytical work has been done on the problem of the unsteady response of separated flow in a cascade, a problem that is of considerable importance in the context of turbomachinery. Though progress has been made in understanding the "dynamic stall" of a single foil (see, for example, Ham 1968), there seems to be a clear need for further research on the unsteady behavior of separated and cavitating flows in cascades. The unsteady lift and moment coefficients are not only valuable in determining the unsteady characteristics of propulsion and lift systems but have also been used to predict the flutter and divergence characteristics of cavitating foils (for example, Brennen et al. 1980).

References

Ackerberg, R.C. (1970). Boundary layer separation at a free streamline. Part 1. Two-dimensional flow. *J. Fluid Mech.*, **44**, 211–225.

Ackerberg, R.C. (1975). The effects of capillarity on free streamline separation. *J. Fluid Mech.*, **70**, 333–352.

Acosta, A.J. (1955). A note on partial cavitation of flat plate hydrofoils. *Calif. Inst. of Tech. Hydro. Lab. Rep. E-19.9.*

Acosta, A.J. (1960). Cavitating flow past a cascade of circular arc hydrofoils. *Calif. Inst. of Tech. Hydro. Lab. Rep. E-79.2.*

Acosta, A.J. (1973). Hydrofoils and hydrofoil craft. *Ann. Rev. Fluid Mech.*, **5**, 161–184.

Acosta, A.J. and Hollander, A. (1959). Remarks on cavitation in turbomachines. *Calif. Inst. of Tech. Eng. Div. Rep. No. 79.3*

Acosta, A.J. and DeLong, R.K. (1971). Experimental investigation of non-steady forces on hydrofoils oscillating in heave. *Proc. IUTAM Symp. on non-steady flow of water at high speeds, Leningrad, USSR*, 95–104.

Ai, D.K. (1966). The wall effect in cavity flows. *ASME J. Basic Eng.*, **88**, 132–138.

Arakeri, V.H. (1975). Viscous effects on the position of cavitation separation from smooth bodies. *J. Fluid Mech.*, **68**, 779–799.

Armstrong, A.H. (1953). Abrupt and smooth separation in plane and axisymmetric flow. *Memo. Arm. Res. Est., G.B., No. 22/63.*

Armstrong, A.H. and Dunham, J.H. (1953). Axisymmetric cavity flow. *Rep. Res. Est., G.B., No. 12/53.*

Armstrong, A.H. and Tadman, K.G. (1954). Axisymmetric cavity flow about ellipsoids. *Proc. Joint Admiralty-U.S.Navy Meeting on Hydroballistics.*

Betz, A. and Petersohn, E. (1931). Application of the theory of free jets. *NACA TM No. 667.*

Biot, M.A. (1942). Some simplified methods in airfoil theory. *J. Aero. Sci.*, **9**, No. 5, 185–190.

Birkhoff, G. and Zarantonello, E.H. (1957). *Jets, wakes, and cavities*. Academic Press.

Brennen, C. (1969a). A numerical solution of axisymmetric cavity flows. *J. Fluid Mech.*, **37**, 671–688.

Brennen, C. (1969b). Some viscous and other real fluid effects in fully developed cavity flows. In *Cavitation State of Knowledge* (eds: J.M. Robertson, G.F. Wislicenus), ASME, N.Y.

Brennen, C. (1970). Some cavitation experiments with dilute polymer solutions. *J. Fluid Mech.*, **44**, 51–63.

Brennen, C.E. and Acosta, A.J. (1973). Theoretical, quasistatic analysis of cavitation compliance in turbopumps. *J. Spacecraft and Rockets*, **10**, No. 3, 175–180.

Brennen, C.E., Oey, K., and Babcock, C.D. (1980). On the leading edge flutter of cavitating hydrofoils. *J. Ship Res.*, **24**, No. 3, 135–146.

Brillouin, M. (1911). Les surfaces de glissement de Helmholtz et la résistance des fluides. *Ann. Chim. Phys.*, **23**, 145–230.

Churchill, R.V. (1948). *Introduction to complex variables and applications*. McGraw-Hill Book Company.

Cohen, H. and Gilbert, R. (1957). Two-dimensional, steady, cavity flow about slender bodies in channels of finite width. *ASME J. Appl. Mech.*, **24**, 170–176.

Cohen, H., Sutherland, C.D., and Tu, Y-O. (1957). Wall effects in cavitating hydrofoil flow. *J. Ship Res.*, **1**, No. 3, 31–39.

Cumberbatch, E. and Wu, T.Y. (1961) Cavity flow past a slender pointed hydrofoil. *J. Fluid Mech.*, **11**, 187–208.

Deshpande, M., Feng, J., and Merkle, C. (1993). Navier-Stokes analysis of 2-D cavity flows. *ASME Cavitation and Multiphase Flow Forum*, **FED-153**, 149–155.

Duller, G.A. (1966). On the linear theory of cascades of supercavitating hydrofoils. *U.K. Nat. Eng. Lab. Rep. No. 218.*

Efros, D.A. (1946). Hydrodynamical theory of two-dimensional flow with cavitation. *Dokl. Akad. Nauk. SSSR*, **51**, 267–270.

Eisenberg, P. and Pond, H.L. (1948). Water tunnel investigations of steady state cavities. *David Taylor Model Basin Rep. No. 668.*

Fabula, A.G. (1962). Thin airfoil theory applied to hydrofoils with a single finite cavity and arbitrary free-streamline detachment. *J. Fluid Mech.*, **12**, 227–240.

Fabula, A.G. (1964). Choked flow about vented or cavitating hydrofoils. *ASME J. Basic Eng.*, **86**, 561–568.

Fage, A. and Johansen, F.C. (1927). On the flow of air behind an inclined flat plate of infinite span. *Proc. Roy. Soc., London, Series A*, **116**, 170–197.

Franc, J.P. and Michel, J.M. (1985). Attached cavitation and the boundary layer: experimental investigation and numerical treatment. *J. Fluid Mech.*, **154**, 63–90.

Furuya, O. and Acosta, A.J. (1973). A note on the calculation of supercavitating hydrofoils with rounded noses. *ASME J. Fluids Eng.*, **95**, 222–228.

Furuya, O. (1974). Supercavitating linear cascades with rounded noses. *ASME J. Basic Eng., Series D*, **96**, 35–42.

Furuya, O. (1975a). Exact supercavitating cascade theory. *ASME J. Fluids Eng.*, **97**, 419–429.

Furuya, O. (1975b). Three-dimensional theory on supercavitating hydrofoils near a free surface. *J. Fluid Mech.*, **71**, 339–359.

Garabedian, P.R. (1956). The mathematical theory of three-dimensional cavities and jets. *Bull. Amer. Math. Soc.*, **62**, 219–235.

Gilbarg, D. (1949). A generalization of the Schwarz-Christoffel transformation. *Proc. U.S. Nat. Acad. Sci.*, **35**, 609–612.

Gilbarg, D. (1960). Jets and cavities. In *Handbuch der Physik*, Springer-Verlag, **9**, 311–445.

Gurevich, M.I. (1961). *Theory of jets in ideal fluids*. Academic Press, N.Y. (1965).

Ham, N.D. (1968). Aerodynamic loading on a two-dimensional airfoil during dynamic stall. *AIAA J.*, **6**, 1927–1934.

Helmholtz, H. (1868). Über diskontinuierliche Flüssigkeitsbewegungen. *Monatsber. Akad. Wiss., Berlin*, **23**, 215–228.

Hsu, C.C. (1972). On flow past a supercavitating cascade of cambered blades. *ASME J. Basic Eng., Series D*, **94**, 163–168.

Hsu, E.-Y. and Perry, B. (1954). Water tunnel experiments on spheres in cavity flow. *Calif. Inst. of Tech. Hydro. Lab. Rep. No. E-24.9*.

Johnson, V.E. (1961). Theoretical and experimental investigation of supercavitating hydrofoils operating near the free surface. *NASA TR R-93*.

Joukowski, N.E. (1890). I. A modification of Kirchhoff's method of determining a two-dimensional motion of a fluid given a constant velocity along an unknown streamline. II. Determination of the motion of a fluid for any condition given on a streamline. *Mat. Sbornik (Rec. Math.)*, **15**, 121–278.

Kelly, H.R. (1967). An extension of the Woods theory for unsteady cavity flows. *ASME J. Basic Eng.*, **89**, 798–806.

Kinnas, S.A. and Fine, N.E. (1990). Non-linear analysis of the flow around partially or supercavitating hydrofoils on a potential based panel method. *Proc. IABEM-90 Symp. Int. Assoc. for Boundary Element Methods, Rome*, 289–300.

Kirchhoff, G. (1869). Zur Theorie freier Flüssigkeitsstrahlen. *Z. reine Angew. Math.*, **70**, 289–298.

Kreisel, G. (1946). Cavitation with finite cavitation numbers. *Admiralty Res. Lab. Rep. R1/H/36*.

Landweber, L. (1951). The axially symmetric potential flow about elongated bodies of revolution. *David Taylor Model Basin Report No. 761*.

Leehey, P. (1971). Supercavitating hydrofoil of finite span. *Proc. IUTAM Symp. on Nonsteady Flow of Water at High Speeds, Leningrad*, 277–298.

Lemonnier, H. and Rowe, A. (1988). Another approach in modelling cavitating flows. *J. Fluid Mech.*, **195**, 557–580.

Levi-Civita, T. (1907). Scie e leggi di resistenzia. *Rend. Circ. Mat. Palermo*, **18**, 1–37.

Martin, M. (1962). Unsteady lift and moment on fully cavitating hydrofoils at zero cavitation number. *J. Ship Res.*, **6**, No. 1, 15–25.

Mimura, Y. (1958). The flow with wake past an oblique plate. *J. Phys. Soc. Japan*, **13**, 1048–1055.

Newman, J.N. (1956). Supercavitating flow past bodies with finite leading edge thickness. *David Taylor Model Basin Rep. No. 1081*.

Nishiyama, T. (1970). Lifting line theory of supercavitating hydrofoil of finite span. *ZAMM*, **50**, 645–653.

Nishiyama, T. and Ota, T. (1971). Linearized potential flow models for hydrofoils in supercavitating flows. *ASME J. Basic Eng.*, **93**, Series D, 550–564.

Numachi, F. (1961). Cavitation tests on hydrofoils designed for accelerating flow cascade: Report 1. *ASME J. Basic Eng.*, **83**, Series D, 637–647.

Numachi, F. (1964). Cavitation tests on hydrofoils designed for accelerating flow cascade: Report 3. *ASME J. Basic Eng.*, **86**, Series D, 543–555.

Parkin, B.R. (1958). Experiments on circular-arc and flat plate hydrofoils. *J. Ship Res.*, **1**, 34–56.

Parkin, B.R. (1959). Linearized theory of cavity flow in two-dimensions. *The RAND Corp. (Calif.) Rep. No. P-1745*.

Parkin, B.R. (1962). Numerical data on hydrofoil reponse to non-steady motions at zero cavitation number. *J. Ship Res.*, **6**, No. 1, 40–42.

Rankine, W.J.M. (1871). On the mathematical theory of stream lines, especially those with four foci and upwards. *Phil. Trans. Roy. Soc.*, 267–306.

Reichardt, H. (1945). The physical laws governing the cavitation bubbles produced behind solids of revolution in a fluid flow. *Kaiser Wilhelm Inst. Hyd. Res., Gottingen, TPA3/TIB.*.

Reichardt, H. and Munzner, H. (1950). Rotationally symmetric source-sink bodies with predominantly constant pressure distributions. *Arm. Res. Est. Trans. No. 1/50.*

Riabouchinsky, D. (1920). On steady fluid motion with a free surface. *Proc. London Math. Soc.*, **19**, 206–215.

Roshko, A. (1954). A new hodograph for free streamline theory. *NACA TN 3168.*

Rouse, H. and McNown, J.M. (1948). Cavitation and pressure distribution: headforms at zero angles of yaw. *Bull. St. Univ. Iowa, Eng., No. 32.*

Sedov, L.I. (1966). *Plane problems in hydrodynamics and aerodynamics (in Russian).* Izdat. "Nauka", Moscow.

Silberman, E. (1959). Experimental studies of supercavitating flow about simple two-dimensional bodies in a jet. *J. Fluid Mech.*, **5**, 337–354.

Sisto, F. (1967). Linearized theory of non-stationary cascades at fully stalled or supercavitating conditions. *Zeitschrift fur Angewandte Mathematik und Mechanik*, **8**, 531–542.

Southwell, R.V. and Vaisey, G. (1946). Fluid motions characterized by free streamlines. *Phil. Trans.*, **240**, 117–161.

Southwell, R.V. (1948). *Relaxation methods in mathematical physics.* Oxford Univ. Press.

Stripling, L.B. and Acosta, A.J. (1962). Cavitation in turbopumps—Part 1. *ASME J. Basic Eng., Series D*, **84**, 326–338.

Struck, H.G. (1970). Discontinuous flows and free streamline solutions for axisymmetric bodies at zero and small angles of attack. *NASA TN D-5634.*

Sutherland, C.D. and Cohen, H. (1958). Finite cavity cascade flow. *Proc. 3rd U.S. Nat. Cong. of Appl. Math.*, 837–845.

Timman, R. (1958) A general linearized theory for cavitating hydrofoils in nonsteady flow. *Proc. 2nd ONR Symp. on Naval Hydrodynamics*, 559–579.

Trefftz, E. (1916). Über die Kontraktion kreisförmiger Flüssigkeits-strahlen. *Z. Math. Phys.*, **64**, 34–61.

Tulin, M.P. (1953). Steady two-dimensional cavity flows about slender bodies. *David Taylor Model Basin Rep. 834.*

Tulin, M.P. (1959). Supercavitating flow past slender delta wings. *J. Ship Res.*, **3**, No. 3, 17–22.

Tulin, M.P. (1964). Supercavitating flows—small perturbation theory. *J. Ship Res.*, **7**, No. 3, 16–37.

Uhlman, J.S. (1978). A partially cavitated hydrofoil of finite span. *ASME J. Fluids Eng.*, **100**, No. 3, 353–354.

Uhlman, J.S. (1987). The surface singularity method applied to partially cavitating hydrofoils. *J. Ship Res.*, **31**, No. 2, 107–124.

Uhlman, J.S. (1989). The surface singularity or boundary integral method applied to supercavitating hydrofoils. *J. Ship Res.*, **33**, No. 1, 16–20.

Villat, H. (1914). Sur la validité des solutions de certains problèmes d'hydrodynamique. *J. Math. Pures Appl.(6)*, **10**, 231–290.

Wade, R.B. (1967). Linearized theory of a partially cavitating cascade of flat plate hydrofoils. *Appl. Sci. Res.*, **17**, 169–188.

Wade, R.B. and Acosta, A.J. (1966). Experimental observations on the flow past a planoconvex hydrofoil. *ASME J. Basic Eng.*, **88**, 273–283.

Wade, R.B. and Acosta, A.J. (1967). Investigation of cavitating cascades. *ASME J. Basic Eng., Series D*, **89**, 693–706.

Wang, D.P. and Wu, T.Y. (1965). General formulation of a perturbation theory for unsteady cavity flows. *ASME J. Basic Eng.*, **87**, 1006–1010.

Widnall, S.E. (1966). Unsteady loads on supercavitating hydrofoils. *J. Ship Res.*, **9**, 107–118.

Woods, L.C. (1957). Aerodynamic forces on an oscillating aerofoil fitted with a spoiler. *Proc. Roy. Soc. London, Series A*, **239**, 328–337.

Woods, L.C. (1951). A new relaxation treatment of flow with axial symmetry. *Quart. J. Mech. Appl. Math.*, **4**, 358–370.

Woods, L.C. (1961). *The theory of subsonic plane flow.* Cambridge Univ. Press.

Woods, L.C. and Buxton, G.H.L. (1966). The theory of cascade of cavitating hydrofoils. *Quart. J. Mech. Appl. Math.*, **19**, 387–402.

Wu, T.Y. (1956). A free streamline theory for two-dimensional fully cavitated hydrofoils. *J. Math. Phys.*, **35**, 236–265.

Wu, T.Y. (1957). A linearized theory for nonsteady cavity flows. *Calif. Inst. of Tech. Eng. Div. Rep. No. 85–6.*

Wu, T.Y. (1962). A wake model for free streamline flow theory, Part 1. Fully and partially developed wake flows and cavity flows past an oblique flat plate. *J. Fluid Mech.*, **13**, 161–181.

Wu, T.Y. (1969). Cavity flow analysis; a review of the state of knowledge. In *Cavitation State of Knowledge* (eds: J.M. Robertson, G.F. Wislicenus), ASME, N.Y.

Wu, T.Y. (1972). Cavity and wake flows. *Ann. Rev. Fluid Mech.*, **4**, 243–284.

Wu, T.Y. and Wang, D.P. (1964a). A wake model for free streamline flow theory, Part 2. Cavity flows past obstacles of arbitrary profile. *J. Fluid Mech.*, **18**, 65–93.

Wu, T.Y. and Wang, D.P. (1964b). An approximate numerical scheme for the theory of cavity flows past obstacles of arbitrary profile. *ASME J. Basic Eng.*, **86**, 556–560.

Wu, T.Y., Whitney, A.K., and Brennen, C. (1971). Cavity-flow wall effect and correction rules. *J. Fluid Mech.*, **49**, 223–256.

Index

Printed in the United States
By Bookmasters